Texts in Philosophy
Volume 15

Knowledge, Value, Evolution

Volume 4
Computing, Philosophy and Cognition
Lorenzo Magnani and Riccardo Dossena, eds.

Volume 5
Causality and Probability in the Sciences
Federica Russo and Jon Williamson, eds.

Volume 6
A Realist Philosophy of Mathematics
Gianluigi Oliveri

Volume 7
Hugh MacColl: An Overview of his Logical Work with Anthology
Shahid Rahman and Juan Redmond

Volume 8
Bruno di Finetti: Radical Probabilist
Maria Carla Galavotti, ed.

Volume 9
Language, Knowledge, and Metaphysics. Proceedings of the First SIFA Graduate Conference
Massimiliano Carrara and Vittorio Morato eds.

Volume 10
The Socratic Tradition. Questioning as Philosophy and as Method
Matti Sintonen, ed.

Volume 11
PhiMSAMP. Philosophy of Mathematics: Sociological Aspects and Mathematical Practice
Benedikt Löwe and Thomas Müller, eds.

Volume 12
Philosophical Perspectives on Mathematical Practice
Bart Van Kerkhove, Jonas De Vuyst and Jean-Paul Van Bendegem, eds.

Volume 13
Beyond Description. Naturalism and Normativity
Marcin Milkowski and Kontrad Talmud-Kaminski, eds.

Volume 14
Corroborations and Criticisms. Forays with the Philosophy of Karl Popper
Ivor Grattan-Guinness

Volume 15
Knowledge, Value, Evolution.
Tomáš Hříbek and Juraj Hvorecký, eds.

Texts in Philosophy Series Editors
Vincent F. Hendriks   vincent@hum.ku.dk
John Symons   jsymons@utep.edu
Dov Gabbay   dov.gabbay@kcl.ac.uk

Knowledge, Value, Evolution

Proceedings of an International Conference
on Cross-Pollinations between
Life Sciences and Philosophy
Prague, November 23-25, 2009

Edited by
Tomáš Hříbek
and
Juraj Hvorecký

© Individual author and College Publications 2011. All rights reserved.

ISBN 978-1-84890-043-1

College Publications
Scientific Director: Dov Gabbay
Managing Director: Jane Spurr
Department of Computer Science
King's College London, Strand, London WC2R 2LS, UK

http://www.collegepublications.co.uk

Original cover design by orchid creative   www.orchidcreative.co.uk
Printed by Lightning Source, Milton Keynes, UK

All rights reserved. No part of this publication may be reproduced, stored in a retrieval system transmitted in any form, or by any means, electronic, mechanical, photocopying, recording or otherw without prior permission, in writing, from the publisher

Acknowledgments

In the first place, we should like to thank all the participants at the conference "Knowledge, Value, Evolution" that we organized in Prague on November 23-25, 2009. In particular, we thank those who developed their conference talks into book chapters that make up this volume. Although we were not able to include here all the papers, we are hoping that the resulting book conveys some of the atmosphere of the mutual learning experience that we enjoyed in Prague. Finally, we appreciate the financial support from the Czech Science Foundation which funded both the conference and the publication of this book with the research grant No. 401/08/0904.

Preface

Tomáš Hříbek and Juraj Hvorecký

The earlier versions of the chapters that make up this volume were originally presented at an international conference "Knowledge, Value, Evolution" that we organized in Prague in November 2009. That was the year of Darwin's double anniversary (200 years since his birth and 150 years since the publication of *The Origin of Species*), and though we certainly wished to join in the celebrations that were going on around the world, our primary goal was not historical. Rather, we wanted to bring together researchers from many different countries who work on topics at the interface of philosophy, the humanities and evolutionary biology. We are proud that we were able to welcome in the capital of the Czech Republic guests not only from the East, Central and South Eastern Europe, Great Britain and Scandinavia, but from as far as the United States, India and Japan. We believe that all of them managed to avoid a merely antiquarian interest, and that the chapters included in this volume give a very comprehensive picture of the work on a Darwinian-inspired epistemology, philosophy of mind, ethics, social philosophy, as well as a more empirical study of cognition and religion. Moreover, the chapters express a whole range of opinions about the influence of Darwinism on philosophy and the humanities. Some of the voices represented in our book are very optimistic, if not downright ecstatic, about the import of evolutionary biology for philosophy and the humanities. Others are cautious, if not sceptical. Either way, all of the contributors to this volume share the opinion that Darwinism represents one of the most important intellectual challenges, and that its topicality is likely to increase in the years to come.

We start with a chapter by one of the veterans of evolutionary philosophy, Franz Wuketits, who usefully surveys his current views about the prospects of a Darwinian epistemology and ethics. If Wuketits is, if we may say so, one of the true believers in a naturalistic, Darwinian philosophy, the author of the following chapter is one of the most outspoken critics of this philosophical programme. Anthony O'Hear expresses deep reservations about the Darwinian metaphysics and epistemology as well as about an allegedly negative impact of Darwin's views on modern social policy. Konrad Talmont-Kaminski then takes issue with O'Hear's diagnosis and defends an unabashedly naturalistic perspective on cognition and ethics. This dialogue is temporarily closed by O'Hear's reply to Talmont-Kaminski.

The next few chapters concentrate on the issue of the scientific status of Darwinism. Lilia Gurova was able to respond very promptly to the ongoing controversy concerning the criticisms of evolutionary biology by the philosopher Jerry Fodor. If most others have been dismissive of Fodor's arguments, Gurova appreciates Fodor's central claim that the role of natural selection in evolutionary explanations is misunderstood. Aviezer Tucker goes back to Darwin's own writings to examine the nature of evolutionary explanations. He argues that Darwinism is an historical science, in the sense that it infers origins of concrete tokens from their effects. Vladimír Havlík, in the next chapter, concurs that Darwinism differs from (say) physics, but he sees the distinction between the two sciences differently from Tucker. Havlík argues that physics or chemistry are confined within their respective domains, while evolutionary biology is an ever-growing body of knowledge, expanding to both lower and higher levels of empirical reality.

The rest of the present volume presents a variety of contemporary views on the two areas of the main theoretical activity within evolutionary philosophy that are already briefly surveyed in Wuketits's chapter—namely, evolutionary epistemology and ethics. We actually construe both of these areas more broadly, thus including contributions to philosophy of mind within the former and the philosophy of the humanities in the latter.

Both Vikram Singh Sirola and Zuzana Parusniková revisit probably the best known version of evolutionary epistemology, namely the theory of Sir Karl Popper. Sirola provides an up-to-date assessment of Popper's account of the growth of knowledge. While by no means uncritical, he regards this project as essentially well-established. By contrast, Parusniková regards Popper's epistemological project as basically incoherent. She discerns elements of dogmatism at the very core of a theory which Popper presented as the very opposite of dogmatism. If sound, her criticism of Popper's epistemology would be devastating. The next two authors turn to more recent attempts to mine the Darwinian insights in the theory of knowledge. Jonathan Knowles comments on the work of Hilary Kornblith who carries on the Quinean project of naturalized epistemology. While coming from a broadly naturalistic and Darwinian tradition himself, Knowles has serious doubts about the success of Korn-blith's work, in particular of his attempt at a scientific vindication of science. While any attempt to justify science assumes that knowledge is valuable, there are sceptics who challenge this very idea. Christos Kyriacou offers a Darwinian answer to these sceptics, arguing that the widely shared intuition that knowledge is valuable is an adaptation. The next four chapters with the applications of evolutionary ideas in the philosophy of mind. Juraj Hvorecký discusses the work of several contemporary philosophers, the so-called teleosemanticists, who turn to Darwinism in an attempt to explain the nature of mental content. Again, Jerry Fodor has been highly critical of these efforts. Hvorecký focuses in particular on the work a leading teleosemanticist, David Papineau, whom he finds conceding too much to Fodor's criticism. Whereas Hvorecký defends the Darwinian project in theory of mind, Tomáš Hříbek appeals to the charge, recently spelled out by Tyler Burge, against the alleged reductionism of naturalistic theories of mind in general, the teleological ones in particular. While he appreciates this critique, Hříbek then goes on to point out that Burge seems to undermine the autonomy of psychology in his own way. Yet others are hopeful that we should be able to tell a continuous Darwinian story that starts at humble causal goings-on and ends with a description of sophisticated intentional behavior. Jaroslav Peregrin offers a sketch of such a story in his chapter. In particular, he believes we could get the normative dimension of intentional behavior by appealing to the Darwinian account. A final chapter to the group of contributions on philosophy of mind is an essay by Ranjan Panda on the concept of emergence, as it has been explored in recent theorizing of Jaegwon Kim and John Searle. Panda explicates difficulties in assigning a causal role to the emergent properties and advocates the Searlean solution to the problem.

The remaining chapters discuss the relevance of the Darwinian ideas in ethics and the study of religion. As far as ethics is concerned, ever since Moore's *Principia Ethica*, attempts at somehow anchoring morality on some facts concerning evolution have been dismissed as so many examples of the naturalistic fallacy. Recently, however, some new naturalists have argued that Darwinian accounts can either debunk or bolster morality. Christoph Schuringa is sceptical of such projects. He examines Nietzsches's older genealogy of morals and wonders whether the Darwinian genealogy does not suffer from inadequacies that can be recognized already in Nietzsche's theory. In the following chapter, Makoto Suzuki concentrates on the debunking projects in the Darwinian meta-ethics—in other words, he tackles the issue of moral scepticism. Suzuki goes on to defend a version of moral realism and objectivism. Wojciech Załuski compares the biological and the sociological accounts of our sense of justice and argues for a mixed approach. The sense of justice is a biological adaptation, but it is not completely innate. In the very last chapter, Slawomir Sztajer surveys theories that look at religion from the point of view of its adaptive value. Sztajer shows how this Darwinian approach changes our view of religion—for example, as religion becomes the topic of a scientific study, the old rivalry between science and religion becomes obsolete.

We hope that our volume offers something for everybody with interests in the current philosophical naturalism in general, and the Darwinian philosophy in particular. We only have one gentle regret. The subtitle of our conference spoke of the "cross-pollinations between life sciences and philosophy." Well, except for one distinguished scientist, all of us who participated at the conference were philosophers, so we could speak of the pollination of philosophy (and the humanities) by evolutionary biology, but not vice versa. It would be great to hear from some researchers in life sciences whether they have something to learn from philosophical reflection on Darwinism. Perhaps we shall be able to find scientists that find philosophy relevant to their concerns and invite them to join in a discussion on another occasion.

Contents

Acknowledgments	v
Preface	vii
Contributors	xiii

1 Evolution, Cognition, and Morality: What Do Epistemology and Moral Philosophy Mean After Darwin?   1
Franz M. Wuketits

2 Darwinian Tensions   17
Anthony O'Hear

3 Evolution, Cognition and Value: the Ingredients for a Naturalist Philosophy   29
Konrad Talmont-Kaminski

4 Reply to Konrad Talmont-Kaminski   49
Anthony O'Hear

5 Fodor vs. Darwin: A Methodological Follow-Up   55
Lilia Gurova

6 Darwin's Inference of Origins   65
Aviezer Tucker

7 The Scientific Status of Darwinism   85
Vladimír Havlík

8 Revisiting Popper's Evolutionary Theory of Knowledge   99
Vikram Singh Sirola

9 Criticism and Dogmatism in Popper's Evolutionary Epistemology 109
Zuzana Parusniková

10 Reciprocal Containment, Naturalized Epistemology and Metaphysical Realism 125
Jonathan Knowles

11 Evolutionary Ruminations on "the Value of Knowledge Intuition" 141
Christos Kyriacou

12 Teleology as a Theory of Meaning 157
Juraj Hvorecký

13 Adaptationism, Deflationism, and Anti-Individualism 167
Tomáš Hříbek

14 Creatures of Norms as Uncanny Niche Constructors 189
Jaroslav Peregrin

15 Searle and Kim on Emergentism 199
Ranjan K. Panda

16 Genealogy, Evolution, and Morality 215
Christoph Schuringa

17 Evolution and Moral Scepticism 227
Makoto Suzuki

18 Evolutionary Origins of the Sense of Justice 257
Wojciech Załuski

19 Evolution, Religion, and the Human Mind 273
Slawomir Sztajer

Contributors

Lilia Gurova, Department of Cognitive Science and Psychology, New Bulgarian University, Sofia, Bulgaria.

Vladimír Havlík, Department of Analytic Philosophy, Institute of Philosophy, Prague, Czech Republic.

Tomáš Hříbek, Department of Analytic Philosophy, Institute of Philosophy, Prague, Czech Republic.

Juraj Hvorecký, Department of Analytic Philosophy, Institute of Philosophy, Prague, Czech Republic.

Jonathan Knowles, Department of Philosophy, Norwegian University of Science and Technology, Trondheim, Norway.

Christos Kyriacou, School of Philosophy, University of Edinburgh, Edinburgh, Great Britain.

Anthony O'Hear, Royal Institute of Philosophy, University of Buckingham, Buckingham, Great Britain.

Ranjan K. Panda, Department of Humanities and Social Sciences, Indian Institute of Technology Bombay, Mumbai, India.

Zuzana Parusniková, Department of Analytic Philosophy, Institute of Philosophy, Prague, Czech Republic.

Jaroslav Peregrin, Department of Logic, Institute of Philosophy, Prague, Czech Republic.

Vikram Singh Sirola, Department of Humanities and Social Science, Indian Institute of Technology Bombay, Mumbai, India.

Christoph Schuringa, Department of Philosophy, London School of Economics, London, Great Britain.

Makoto Suzuki, Department of Philosophy, Keiko University, Mita, Japan.

Slawomir Sztajer, Institute of Philosophy, Adam Mickiewicz University, Poznan, Poland.

Konrad Talmont-Kaminski, Institute of Philosophy, Marie Curie-Sklodowska University, Lublin, Poland.

Aviezer Tucker, CEVRO Institute, Prague, Czech Republic.

Franz M. Wuketits, Institute of Philosophy of Science, University of Vienna, and Konrad Lorenz Institute for Evolution and Cognition Research, Altenberg, Austria.

Wojciech Załuski, Department of Legal Philosophy and Legal Ethics, Jagiellonian University, Krakow, Poland.

1

Evolution, Cognition, and Morality: What Do Epistemology and Moral Philosophy Mean After Darwin?

Franz M. Wuketits

> He who understands baboon would do more towards metaphysics than Locke.
>
> (Charles Darwin)

> Of all histories the history of ideas is the most difficult and elusive. Unlike things, ideas cannot be handled, weighed, and measured
>
> (John C. Greene)

1 Introduction

In the concluding chapter of Darwin's *Origin of Species* we find the frequently quoted short—and somewhat cryptic—statement that "much light will be thrown on the origin of man and his history" (Darwin 1958 [1859], 449). However, it took Darwin twelve more years to publish his *Descent of Man* (1871), a seminal work that in fact threw much light on humans and their history. In these years some other naturalists—most prominently among them Thomas H. Huxley in England and Ernst Haeckel in Germany—gave evidence to the assertion that humans are results of evolution and came up from ape-like creatures. For different reasons, Darwin was cautious and hesitant, but there is no doubt that he had speculated on the origin of humankind long before the publication of *Descent of Man* and even when he was preparing the *Origin*. His *Beagle Diary* (Darwin 1988) includes a bulk of observations concerning the behavior and customs of people in different regions of the earth, particularly in South America (see also Richards 1987), so that we can be sure that when he stated "much light will be thrown" he already knew very well what he was announcing.

Descent of Man is a brilliant exposition of Darwin's conclusions regarding the origin and evolution of humans. However, it does not just offer evidence for human biological evolution, but—what is more—includes full-length chapters on human mental, social, and moral capacities and their evolutionary origins. Thus, the book is an outline of some fundamental principles of an evolutionary theory of cognition and knowledge as well as an evolutionary theory of morality. Its relevance for philosophy is beyond any doubt. Actually, if one takes Darwin and the theory of evolution really seriously, philosophy can never again be what it had been before (Oeser 1987). Yet many philosophers—especially those standing in the "idealist" tradition—tend to disregard evolutionary thinking which, for some of them, appears to be a heresy. If our species is—and it definitively is—a result of evolution by natural selection, then of course all its capacities addressed by philosophers need to be described and explained in evolutionary terms. What follows, is an overall assessment of the meaning of the Darwinian worldview in epistemology and moral philosophy or ethics. This is an enormous topic. I have to omit some of its particular aspects and to present other aspects in a nutshell; however, I hope that the basic assumptions and implications of "Darwinian philosophy" will become visible and that this chapter can stimulate further discussions.

2 Philosophy Naturalized

If, according to Darwin, we regard (human) mental and moral phenomena as *natural* phenomena, then we arrive at the conclusion that at least some philosophical disciplines (epistemology, ethics, philosophy of language, and others) have to be naturalized. In fact, Darwin already "took" what later was called the *naturalistic turn* (see Callebaut 1993). He closed the long supposed gap between humans and other animals and claimed that the differences between *Homo sapiens* and other species are just gradual ones (see Darwin 1871, 1872). His very message could be put briefly and accurately as follows: "We are biological and our souls cannot fly free" (Wilson 1979, 1). It should be noted that most of Darwin's contemporaries—even some evolutionists (above all Alfred R. Wallace)—continued to look at (human) mental phenomena as something beyond the reach of evolutionary theory (see Richards 1987). They considered the mind to be a supernatural entity unique to humans and, vice versa, tried to substantiate the idea of the uniqueness of human beings by referring to their mental abilities.

Darwin and evolutionary theory do not leave any space for the assumption of unnatural or supernatural entities. Our knowledge, our thinking and reasoning, our language as well as our moral sentiments are to be regarded as particular brain functions. The brain is a biological organ that, like any other organ, evolved by natural selection. Therefore, as a matter of principle, the theory of evolution offers the *ultimate* explanation of all mental abilities.[1] This does not mean to "dephilosophize" problems of the mind, but to handle them in an even broader perspective. If there is nothing unnatural about mind, then it must be based on natural history—and then all its expressions (including moral sentiments) must have natural groundings. My aim in the present paper is to elucidate the meaning of evolutionary epistemology and evolutionary ethics in this context. I am aware that this is a rather sketchy treatment, and I feel that some readers will find my exposition somehow dogmatic. If this is really the case, I have one—as I think—quite a good apology: Dogmatism of some kind can stimulate opposition and help advance ideas.

3 Evolutionary Epistemology

"An evolutionary epistemology would be at minimum an epistemology taking cognizance of and being compatible with man's status as a product of biological and social evolution." Thus Campbell (1974, 413), in an essay regarded as a classic of this type of epistemology. Campbell also claimed that evolution itself is a cognition process and that the natural-selection paradigm can be extended to epistemic activities like learning, thinking, and scientific knowledge processing.

Evolutionary epistemology, in other words, is an attempt to describe, reconstruct, and explain cognitive phenomena in terms of evolutionary theory. It is based on the assumption that cognition of any type—in humans and other animals—is a biological phenomenon resulting from evolution by natural selection. Moreover, from the point of view of this epistemology, cognitive mechanisms serve life-preserving functions and thus increase the fitness of a living being (see Wuketits 1997). Yet, what seems trivial to the biologist, has challenged– and sometimes "taken aback"–many philosophers. However, evolutionary epistemology–as a *naturalized epistemology* (Oeser 1983; Quine 1971)–has a quite long and venerable tradition. Among its early advocates we find the philosopher and social scientist Herbert Spencer, the naturalist Ernst Haeckel, the

---

[1] Here I refer to the useful proximate/ultimate distinction (Mayr 1961, 1993). A biologist who wants to explain, for example, why a dog is barking, can find a proximate explanation by some external stimuli that make the dog aggressive. But such an explanation does not answer the questions, why dogs bark at all. Only some profound knowledge of the origins and evolutionary pathways of dogs will help to understand this behavioral trait common to all races of this species and their ancestors, the wolves. Likewise, the proximate causes of any particular human mental state can be found in some immediate factors (e.g., constrained by social and/or cultural circumstances) whereas its ultimate causes are deeply rooted in the evolution of human mind or, better to say, the evolution of the (human) brain.

physicists and philosophers Ernst Mach and Ludwig Boltzmann, and many others (for a historical survey see Campbell 1974 and Wuketits 1990).[2]

In the German speaking countries evolutionary epistemologists soon involved themselves in an attempt to interpret—and to "remodel"—the epistemology of Immanuel Kant in evolutionary terms (Lorenz 1941, 1977; see Wuketits 1987, 1990 for discussion). Needless to say that Kantian epistemology has been most influential on the European continent for two centuries. Kant's distinction between *a priori* and *a posteriori* was—and still is—to be considered as a significant epistemological step towards a synthesis going beyond *empiricism* and *rationalism*. However, what Kant regarded as *a priori*, given categories and forms of intuition, can be interpreted as the *a posteriori* of evolution: Human reason—and with it any knowledge *a priori*—has evolved, is a result of "evolutionary learning programs" that were supported and stabilized by natural selection. In other words: "The prerequisites of human thinking, though *a priori* for each individual in the sense of Kant, are *a posteriori* for the chain of his pedigree" (Mohr 1977, 198). Hence, evolutionary epistemology has been the "dynamization" of the *a priori* and of the whole corpus of Kantian epistemology.

But neither in German speaking countries nor elsewhere have advocates of evolutionary epistemology confined themselves to an endorsement of Kantian epistemology in evolutionary terms. The program of evolutionary epistemology has been much broader, inspired by ideas of several disciplines (see, e. g., Bradie 1986, Callebaut and Pinxten 1987, Hahlweg and Hooker 1989, Radnitzky and Bartley 1987, Riedl and Wuketits 1987, Wuketits 1984, 1990). In fact, there have been two somewhat distinct programs of evolutionary epistemology. First is the attempt to account for cognitive mechanisms in animals and humans by the extension of the biological theory of evolution to those structures and functions of living beings that are the biological substrates of cognition (brain, nervous systems, sense organs). This is a *natural history* of cognition (knowledge) founded on results in behavioral sciences, sensory physiology, evolutionary biology, and so on (Lorenz 1977). Second, more specifically, is the attempt to explain *human* knowledge (ideas, scientific theories) in terms of evolution; that means, using evolutionary models in the sphere of (human) culture and its products (see Oeser 1988; Popper 1972). However, both programs are interrelated. Once the meaning of evolutionary theory for studying cognition/ knowledge was grasped, it became apparent that evolutionary models can in fact be extended to *all* types of cognitive phenomena, however specific.

---

[2] However, the term "evolutionary epistemology" was coined only in the 1950s, according to my knowledge by Donald Campbell.

Both programs of evolutionary epistemology are underlined by the rejection of the empiricist (sensualist) doctrine of *tabula rasa*. As Popper (1972, 71) accurately stated:

> The *tabula rasa* theory is absurd: At every stage of the evolution of life and of the development of an organism, we have to assume the existence of some knowledge in the form of dispositions and expectations. Accordingly, *the growth of all knowledge consists in the modification of previous knowledge*–either its alteration or its large-scale rejection. Knowledge never begins from nothing, but always from some background knowledge.

Consequently, evolutionary epistemology stands in stark contrast to behaviorism and any doctrine that "reduces" cognition and knowledge processes to individual learning. This does not mean that advocates of evolutionary epistemology disregard the importance of learning. It rather means that they assume "innate teaching mechanisms" as phylogenetically acquired programs. As Lorenz (1977, 89) said:

> Unless one believes in supernatural factors, such as a pre-established harmony between the organism and its environment, one has to postulate the existence of innate teaching mechanisms in order to explain why the majority of learning processes serve to enhance the organism's fitness for survival.

Clearly, evolutionary epistemologists do not believe in supernatural entities and give credit to a naturalist view of the mind, while at the same time they strongly oppose the dualist conception of mind and body as two different (separate) categories.[3] They do not ontologically reduce the mind to the matter–Lorenz (1977), for one, argued that the (human) mind is to be regarded as an emergent property of life–but they claim that mental phenomena can never be sufficiently explained without resort to their biological substrate. Similar or rather the same arguments can be found in sociobiology (see, e.g., Wilson 1979, 1987).

Yet another basic assumption of evolutionary epistemology, which connects its two programs, has to be mentioned: Evolution is a cognition process (see, e. g., Campbell 1974; Lorenz 1977, Wuketits 1990). More precisely, one should say that evolution is a cognition-gaining process. However, this claim has been frequently misunderstood. In what way is evolution actually a cognition or cognition-gaining process? To put it briefly, we can state that all organisms accumulate information about at least some properties of their respective environment and generally can be characterized as information-processing systems;

---

[3] Surprisingly, Karl Popper published together with John C. Eccles–who was probably the last dualist among neurobiologists–an argument for interactionism that is a kind of dualism (see Popper and Eccles 1977).

their capacity to gather information for the sake of survival has been shaped by natural selection. Thus, life is cognition (Heschl 1990), in that sense that only those organisms that are able to process relevant information about their environment are able to perform other elementary life functions (particularly reproduction). Sure, when a lion is not well-informed about its possible prey, about a possible mate, and so on, it will hardly survive.

This brings us to the problem of *realism*. From the point of view of evolutionary epistemology the concept of "reality" is a pragmatic one. Any belief in a "world-in-itself" or entities beyond the perceivable world appears to be obsolete. At the same time, we have no reason to believe that our perception yields a "true" picture of all aspects of reality. Our perceptual apparatus—like that of other species—was developed and shaped just for survival's sake in its specific *cognitive niche* or *mesocosm* (Vollmer 1975). Jakob von Uexküll, though he was not an evolutionist, applied the useful term *ambient* (*Umwelt*) to indicate that any species exists in its specific world and that the range of perception varies from one species to another (Uexküll and Kriszat 1956): Different species perceive different parts of reality, since they are adapted to and live in changing environmental conditions; thus, they generate different "world views." With respect to our own species Lorenz (1977, 7) gave the following explication:

> What we experience is indeed a real image or reality—albeit an extremely simple one, only just sufficing for our own practical purposes; we have developed "organs" only for those aspects of reality of which, in the interest of survival, it was imperative for our species to take account, so that selection pressure produced this particular cognitive apparatus [...] Yet what little our sense organs and nervous system have permitted us to learn has proved its value over endless years of experience, and we may trust it—as far as it goes.

There is a nice metaphor by George G. Simpson that "the monkey who did not have a realistic perception of the tree branch he jumped for was soon a dead monkey—and therefore did not become one of our ancestors" (Simpson 1963, 98). Are monkeys "realists"?

The point here is: The monkey is not forced to *know* what we humans mean when we say "tree branch," nor is he compelled to reflect on the very nature of tree branches (and other objects); what is imperative for the monkey is to recognize that there is something "out there," to which he has to react adequately, i.e., for the sake of his own survival. Neither monkeys, nor other animals (including humans) simply represent the outer world—they also *interpret* what is "out there" according to their own requirements (see also Riegler 2006). They are neither "realists," nor "constructivists," they simply follow the phylogenetically acquired programs that have been supported by natural selection and give them the advice to act concurrently. "Sight of prey or sound of predator must be suitably processed or interpreted to result in use of teeth or use of feet accordingly" (Clark 1986, 151). That's it.

In sum, the most important philosophical implications of evolutionary epistemology are the following (see Wuketits 2006).

(1) The notion of an "essence" or "form"—in the sense of Plato—is definitely obsolete. Anyone who takes evolutionary thinking seriously cannot believe in something beyond each object of the world. The theory of evolution relies on variation and, hence, contradicts typological thinking or essentialism (see also Mayr 2000).

(2) The notion of the *absolute*—here, particularly, "absolute knowledge" and "absolute truth"—is no longer tenable. "Truth" is a human construction and does not exist as a real entity. "True," in a strict biological sense, is what helps survival.

(3) There is no reason to believe in the *unknowable* (and thus to assume the existence of an unknowable world-in-itself). There *is* a real world which we, at least partly, can perceive and, moreover, reconstruct by the means of scientific methods. If there is anything that is indeed unknowable on principle grounds, then we should not worry—since we cannot know it, it will neither help nor disturb us.

For brevity's sake, I must refrain here from bringing more details or paying attention to the critics. However, at least in German speaking countries evolutionary epistemology is meanwhile so well established, that it has been comprehensively presented in a students' textbook (Irrgang 2001). (Which, of course, does not mean that it is accepted by all philosophers.)

Yet, evolutionary epistemology is but one among several attempts to understand human mental capacities as natural phenomena. Disciplines like neurobiology, paleo-neurology, cognitive ethology, and others offer momentous data and perspectives in this respect.[4] But as far as I can see, a *natural history of mind* is still an unfinished synthesis; it is important, in the spirit of a truly interdisciplinary account, to bring together different approaches and conceptions. What I do not believe is that dealing with artificial intelligence will be of much help. It is because of its detachment from biology. As Bunge (1980, 63) accurately stated: "If we wish to understand mind we had better study animals rather than machines." With other animals we have a common descent—with computers we

---

[4] Let me also hint at an "evolutionary theory of emotions" that gives some insight into the "irrational" underpinnings of cognition/knowledge (see Wimmer and Ciompi 2005) and plays a significant role in the search for an understanding of the roots of morality.

do not (rather they are our own creations). Remember also Darwin's statement quoted at the beginning of this chapter.

4 Evolutionary Ethics

A definition of evolutionary ethics can be given analogously to Campbell's definition of evolutionary epistemology: An evolutionary ethics would be, at a minimum, an ethics taking cognizance of and being compatible with the status of humans as results of biological and social evolution. The starting point of any evolutionary attempt to describe and explain morality is that all moral sentiments, of whatever type, have their groundings in biological and social evolution and have been constrained by natural selection (see, e. g., MacDonald 2001; Mohr 1987; Richards 1986; Ridley 1997; Ruse 1986; Wilson 1979; Wuketits 1993, 1995, 1999). Morality is nothing non-natural; it follows from specific requirements in human life and can be regarded as a sophistication of "altruistic programs" that are deeply rooted in the evolution of our species. Moreover, we can state that

(1) evolution of morality is just one—albeit very important—aspect of human evolution;

(2) moral behavior is constrained by the requirements of our lives as a social species, whose members (like the members of all other species) have a particular interest in their survival;

(3) like in any other social species, *cooperation* and *reciprocal altruism* play a crucial role in human social life and can be regarded as the very essence of moral behavior;

(4) humans have invented "good" and "evil" ("right" and "wrong"); by following, one way or another, what they call "good" ("right") in a moral sense, they follow mainly altruistic principles.

Taking for granted that morality is indeed just a sophistication of altruistic strategies that are widely spread in the animal kingdom (particularly among the social species), we arrive at the conclusion that our moral codes follow natural principles that were shaped by selection long before anybody rationally reflected on what we ought to do (or not to do).

Darwin (1871) referred to *social instincts* that he attributed chiefly to natural selection. He appreciated the advantage of sociality and anticipated some basic ideas of modern sociobiology: The expenses of social life at the individual's level, i.e., the necessities to take care of others and to help them, are rewarded by some security and pleasure—and, in the last instance, by individual reproductive success. Moreover, Darwin found indications of a more or less inevitable improvement of the social instincts and gave expression to his hope that in future or social instincts will expand, so that our initially limited sympathy will encompass more and more individuals of our own species and, in last instance, also members of nonhuman species.[5]

Anyway, moral conducts are not imposed on us by a Deity, but are our own creations. Their very purpose is to make our social life possible. Any idealistic conception of humans that does not pay sufficient attention to what humans actually want and desire and which kind of behavior in others they cherish or hate will tempt us to build castles in the air. A *realistic* assessment of moral behavior is needed. Moral reasoning has to go along a profound understanding of the particular conditions of human life. "Moral reasoning matters, but it matters primarily in social contexts in which people try to influence each other and reach consensus with friends and allies" (Capó et al. 2006, 134).[6] Hence, ethical theory needs an empirical basis or, to put it the other way round, ethics can be established as an applied science (Ruse and Wilson 1986). This brings us conveniently to one—if not the most—important conclusion of evolutionary ethics: the relativity of norms and values.

Unless one believes in some "higher moral authority" there is no reason to assume *absolute* norms and values. In fact, while morality is universal in humans—I do not want to speculate here about some behavioral traits especially in chimpanzees that, in a sense, maybe demonstrate the beginnings of morality—it is obvious that concrete norms and values vary from one society (culture) to another. Also, as can be seen from history, norms and values change; we follow today moral standards that differ in some ways from those of, e.g., the ancient Romans. Finally, even during an individual's life the understanding of what is morally right or wrong can change, for we have to take into account "that in ecological and biosocial contexts the individual experience of norms and values strongly varies in situation-specific ways" (Geiger 1992, 319). Certainly, a starving person will tend to circumvent or even to violate moral norms, while the same person, if he or she enjoys a pleasant life, will possibly agree to those norms.

Unfortunately, as the biological theory of evolution generally has been resisted by many sociologists and researchers in some other fields in the study of humans (see Greenwood 1984), there has been resistance to evolutionary ethics in particular, mainly among philosophers (see Farber 1994). This is due

---

[5] To be sure, Darwin was definitely not a Social Darwinist. Rather, one could argue that he was—as I pointed out elsewhere (Wuketits 2009a, b)—a social romanticist. Inspired by the ideas of Enlightenment, he believed in a moral improvement of humankind and embraced the vision of humanity without oppression, racism and discrimination.

[6] We find similar ideas, although not grounded on evolutionary theory, in the writings of Moritz Schlick, one of the central figures of the Vienna Circle (see Leinfellner 1985).

to the widely spread conviction that bringing together evolutionary thinking and moral reasoning means to commit the *naturalistic fallacy*, i.e., to ignore the distinction between *is* and *ought* (fact/value distinction). I hasten to say that evolution, indeed, does not tell us, how we have to act, and that it does not entail any moral standards. Thomas H. Huxley, an ardent defender of Darwin's views ("Darwin's bulldog") even claimed that moral systems have to work *against* evolution (see Williams 1988). This was wrong, because it presupposes that evolution is morally bad. Evolution is morally neutral and humans, as its products, are neither bad nor good by nature. However, ignoring our evolutionary origins means to commit a "counter-naturalistic fallacy" (Simpson 1969). As I argued elsewhere (Wuketits 1993, 1999), moral systems are evolutionary systems in the sense that there are certain facts in the external world that influence the development of particular values and norms. For example, a human population living in the desert will connect certain values with water and thus develop moral codes of the kind "Don't waste water!" (In our Western societies water is obviously not highly valued—simply because it seems to be an unlimited resource.)

Unless one believes in unnatural roots of morality, one has to admit that we humans have our preferences and aversions and that, then, "the empirical preference data are the only real bases for the abstract value order" (Leinfellner 1979, 305). Hence, it can be said that we

- evaluate certain facts of the external world, and
- derive (moral) norms from this evaluation.

When doing so, we usually do not need any reference to some "higher" moral principles. As was already stated, reciprocal altruism has been the motor of our social evolution. If our ancestors had only cheated or even killed each others, then they would not have become ancestors of anyone—and we would not be around any longer. From this point of view Thomas Hobbes' *bellum omnium contra omnes*, a war of all against all, was based on a misconception of human nature. Humans have never been angels, that much is clear, but there is a good reason to believe that natural selection has supported reciprocal altruism for the sake of survival. Sure, we humans, like other animals, are egoists, but as a social species we, like other social animals, depend on members of our own species. Personal utility and altruism appear as closely related. At all stages of our evolutionary development we have to assume some cooperation and reciprocal help, that is to say a kind of "moral minimalism" or "minimal morality." Moral philosophers should feel challenged by the possibility to give Kant's categorical imperative—and similar such imperatives—a sound evolutionary fundament. As was stated previously, there are not objective moral norms or values. But it seems obvious that humans—no matter which society or culture they belong to—are endowed with the same emotional and affective dispositions and usually entertain a strong desire to survive. Couldn't this be the basis for reflections concerning some general views of the "good life" in a moral sense?

Critics of evolutionary ethics—and, especially, those who tend to idealize human moral behavior—seem to have in mind only the negative applications of naturalism; as if evolution had brought forth only immoral behavior, and as if only pure egoism and competitive behavior were standards for an evolutionary approach to morality and ethics. On the contrary, we can apply our knowledge about our own evolutionary history and our biologically constrained preferences in a positive sense. Humans will—so much can be maintained—have a better life, and tend to behave morally, if they can find pleasure and satisfaction in their own lives, if they are not restricted too much (by abstract moral and/or legal imperatives), and if they are prepared to share their happiness with others. In earlier papers (Wuketits 1993, 2009b) I argued that our moral systems should be constructed in such a way that they permit a pleasant life, simply because happy people will be probably better prepared to take care of others, to help and to support others—and to extend their sympathies.

For now, evolutionary ethics has not received too much attention by moral philosophers, at least not as much as it deserves. On the other side, it does find some support in philosophical tradition. A possible precursor is David Hume. There is a remarkable similarity between Hume and Darwin, for the Scottish empiricist had already made morality a function of nature and his assumption of general feeling or sentiment—to which he attributed our ability to cooperate—resembles Darwin's "social instincts" (see Hume 1975).

It is worthwhile to pursue the evolutionary approach to an understanding of morality further. Evolution does not give us an advice regarding what we ought to do, but helps us to answer the question, why we behave like human beings. There are no moral rules in nature—but there is much nature in our moral behavior (Vogel 1989; see also Richards 1986 for similar arguments). We should not try to derive norms/values from nature, but it is our nature that determines us, in some way, to give preference to certain norms/values and to refuse others. To detach our own nature from moral discourse—or to "moralize" without considering our natural constraints—means to build "ethical ivory towers."

5 Conclusion

What do epistemology and moral philosophy or ethics mean after Darwin? Sure, their fundamental questions and problems have not changed, but the approaches to the questions and problems differ from age-old and venerable traditions in philosophy. I have particularly in mind the idealistic tradition that is rooted in Plato's typological thinking and essentialism, which has endured and is still influencing the works of many philosophers. In contrast to this the evolutionary worldview includes variation, dynamics, and complexity; and it is firmly grounded in an historical conception of nature including humans and their activities. As Julian Huxley—the grandson of Thomas H. Huxley and himself an eminent evolutionist—once remarked:

> All phenomena have a historical aspect. From the condensation of nebulae to the development of the infant in the womb, from the formation of the earth as a planet to the making of a political decision, they are all processes in time; and they are all interrelated as partial processes within the single universal process of reality. All reality, in fact, *is* evolution, in the perfectly proper sense that it is a one-way process in time; unitary; continuous; irreversible; self-transforming; and generating variety and novelty during its transformations. (Huxley 1953, 2)

If philosophy—here particularly epistemology and ethics—is to make progress it can no longer disregard human nature and its (evolutionary) history.

After Darwin the invocation of any supernatural entities has become obsolete. This, I trust, is a stance common to most philosophers today. However, many philosophers are obviously not prepared to take really seriously what an evolutionary view of life actually means. This view not only opens new perspectives for theorizing in epistemology, ethics, and other "classical" philosophical disciplines—from philosophy of nature to philosophy of language—but it is also apt to give some fresh impetus to a secular view of humans with crucial "practical" implications concerning our place in nature and our future possibilities in a constantly changing world. As Darwin realized, the knowledge of our very origins and our evolutionary development will improve our self-knowledge. It is, therefore, about time to bring his ideas to philosophy. Instead of ignoring or opposing the Darwinian views philosophers could profit from them and thus help their own discipline to attract more attention as a critical examination of the *conditio humana*.

References

Bradie, M. (1986). Assessing Evolutionary Epistemology. *Biology and Philosophy* 1: 401-450.

Bunge, M. (1980). *The Mind-Body Problem: A Psychobiological Approach*. Oxford: Pergamon Press.

Callebaut, W. (1993). *Taking the Naturalistic Turn or How Real Philosophy of Science is Done*. Chicago: The University of Chicago Press.

_____ and Pinxten, R. (eds.) (1987). *Evolutionary Epistemology: A Multiparadigm Program*. Dordrecht: Reidel.

Campbell, D. T. (1974). Evolutionary Epistemology. In *The Philosophy of Karl Popper I*. P. A. Schilpp (ed.). LaSalle, Ill.: Open Court, 413-463.

Capó, M., Nadal, M., and Cela-Conde, C. J. (2006). Moral Consilience. *Biological Theory* 1: 133-135.

Clark, A. J. (1986). Evolutionary Epistemology and the Scientific Method. *Philosophica* 37: 151-162.

Darwin, Ch. (1859 [1958]). *The Origin of Species by Means of Natural Selection*. New York: New American Library.

―――― (1871). *The Descent of Man and Selection in Relation to Sex*. London: Murray.

―――― (1872). *The Expression of the Emotions in Man and Animals*. London: Murray.

―――― (1988). *Charles Darwin's Beagle Diary*. Ed. by R. Keynes. Cambridge: Cambridge University Press.

Farber, P. L. (1994). *The Temptations of Evolutionary Ethics*. Berkeley: University of California Press.

Geiger, G. (1992). Why There Are No Objective Values: A Critique of Ethical Intuitionism From an Evolutionary Point of View. *Biology and Philosophy* 7: 315-330.

Gontier, N., J. P. Van Bendegem and D. Aerts (eds.) (2006). *Evolutionary Epistemology, Language and Culture: A Non-Adaptationist, Systems Theoretical Approach*. Dordrecht: Springer.

Greenwood, D. J. (1984). *The Taming of Evolution: The Persistence of Nonevolutionary Views in the Study of Humans*. London: Cornell University Press.

Hahlweg, K. and Hooker, C. A. (eds.) (1989). *Issues in Evolutionary Epistemology*. Albany, N.Y.: SUNY Press.

Heschl, A. (1990). $L = C$ A Simple Equation with Astonishing Consequences. *Journal of Theoretical Biology* 145: 13-40.

Hume, D. (1975). *Enquiries Concerning Human Understanding and Concerning the Principles of Morals*. Ed. by P. H. Nidditch. Oxford: Oxford University Press.

Huxley, J. (1953). *Evolution in Action*. New York: Harper & Brothers.

Irrgang, B. (2001). *Lehrbuch der Evolutionären Erkenntnistheorie*. Munich: Reinhardt.

Leinfellner, W. (1985). A Reconstruction of Schlick's Psycho-Sociological Ethics. *Synthese* 64: 317-349.

―――― (1989). The Naturalistic versus the Intuitionistic Schools of Values. In *Expected Utility and the Allais Paradox*. M. Allais and O. Hagen (eds.). Dordrecht: Reidel, 303-332.

Lorenz, K. (1941). Kants Lehre vom Apriorischen im Lichte gegenwärtiger Biologie. *Blätter für Deutsche Philosophie* 15: 94-125.

―――― (1977). *Behind the Mirror: A Search for a Natural History of Human Knowledge*. London: Methuen.

MacDonald, C. (2001). Evolutionary Ethics: Value, Psychology, Strategy and Conventions. *Evolution and Cognition* 7 (1): 98-105.

Mayr, E. (1961). Cause and Effect in Biology. *Science* 134: 1501-1506.

―――― (1993). Proximate and Ultimate Causations. *Biology and Philosophy* 8: 93-94.

―――― (2000). Darwin's Influence on Modern Thought. *Scientific American* 283 (1): 78-83.

Mohr, H. (1977). *Lectures on Structure and Significance of Science*. New York: Springer.

_____ (1987). *Natur und Moral. Ethik in der Biologie.* Darmstadt: Wissenschaftliche Buchgesellschaft.

Oeser, E. (1983). Naturalisierte Erkenntnistheorie und Methodendynamik. In *Wahrheit und Wirklichkeit.* P. Kampits, G. Pöltner and H. Vetter (eds.). Berlin: Duncker & Humblot, 87-99.

_____ (1987). *Psychozoikum. Evolution und Mechanismus der menschlichen Erkenntnisfähigkeit.* Berlin: Parey.

_____ (1988). *Das Abenteuer der kollektiven Vernunft: Evolution und Involution der Wissenschaft.* Berlin: Parey.

Popper, K. R. (1972). *Objective Knowledge: An Evolutionary Approach.* Oxford: Clarendon Press.

_____ and Eccles, J. C. (1977). *The Self and Its Brain: An Argument for Interactionism.* New York: Springer.

Quine, W. V. O. (1971). *Ontological Relativity and Other Essays.* New York: Columbia University Press.

Radnitzky, G. and Bartley, W. W., (eds.) (1987). *Evolutionary Epistemology, Theory of Rationality, and the Sociology of Knowledge.* LaSalle: Open Court.

Richards, R. (1986). A Defense of Evolutionary Ethics. *Biology und Philosophy* 1: 265-293.

Richards, R. (1987). *Darwin and the Emergence of Evolutionary Theories of Mind and Behavior.* Chicago: The University of Chicago Press.

Ridley, M. (1997). *The Origins of Virtue.* London: Penguin Books.

Riedl, R. and Wuketits, F. M. (eds.) (1987). *Die Evolutionäre Erkenntnistheorie: Bedingungen, Lösungen, Kontroversen.* Berlin: Parey.

Riegler, A. (2006). Like Cats and Dogs: Radical Constructivism and Evolutionary Epistemology. In Gontier, Van Bendegem and D. Aerts (2006), 47-65.

Ruse, M. (1986). *Taking Darwin Seriously: A Naturalistic Approach to Philosophy.* Oxford: Basil Blackwell.

_____ and Wilson, E. O. (1986). Ethics as Applied Science. *Philosophy* 61: 173-192.

Simpson, G. G. (1963). *This View of Life: The World of an Evolutionist.* New York: Harcourt, Brace and World.

_____ (1969). *Biology and Man.* New York: Harcourt, Brace & World.

Uexküll, J. v. and Kriszat, G. (1956). *Streifzüge durch die Umwelten von Tieren und Menschen.* Hamburg: Rowohlt.

Vogel, Ch. (1989). *Vom Töten zum Mord: Das wirklich Böse in der Evolutionsgeschichte.* Munich: Hanser.

Vollmer, G. (1975). *Evolutionäre Erkenntnistheorie.* Stuttgart: Hirzel.

Williams, G. C. (1988). Huxley's Evolution and Ethics in Sociobiological Perspectives. *Zygon* 23: 383-407.

Wilson, E. O. (1979). *On Human Nature.* New York: Bantam Books.

_____ (1987). The Evolutionary Origin of Mind. *The Personalist Forum* 3 (1): 11-18.

Wimmer, M. and Ciompi, L., (eds.) (2005). *Emotion, Kognition, Evolution: Biologische, Psychologische, soziodynamische und philosophische Aspekte.* Fürth: Filander.

Wuketits, F. M., (ed.) (1984). *Concepts and Approaches in Evolutionary Epistemology: Towards an Evolutionary Theory of Knowledge.* Dordrecht: Reidel.

_____ (1987). Hat die Biologie Kant missverstanden? Evolutionäre Erkenntnistheorie und "Kantianismus." In *Transzendentale oder evolutionäre Erkenntnistheorie?* W. Lütterfelds (ed.). Darmstadt: Wissenschaftliche Buchgesellschaft, 33-50.

_____ (1990). *Evolutionary Epistemology and Its Implications for Humankind.* Albany, N.Y.: SUNY Press.

_____ (1993). Moral Systems as Evolutionary Systems: Taking Evolutionary Ethics Seriously. *Journal of Social and Evolutionary Systems* 16: 251-271.

_____ (1995). Biosocial Determinants in Moral Behavior: An Evolutionary Approach. *HOMO* 46 (2): 113-124.

_____ (1997). Evolution, Cognition, and Survival: Evolutionary Epistemology and Derivative Topics. *World Futures* 51: 47-93.

_____ (1999). *Warum uns das Böse fasziniert: Die Natur des Bösen und die Illusionen der Moral.* Stuttgart: Hirzel.

_____ (2006). Evolutionary Epistemology: The Non-adaptationist Approach. In Gontier, Bendegem and Aerts (2006), 33-46.

_____ (2009a). Der sanfte Darwin. *Psychologie heute* 36 (3): 60-63.

_____ (2009b). Charles Darwin and Modern Moral Philosophy. *Ludus Vitalis* 17/32: 395-404.

2

Darwinian Tensions

Anthony O'Hear

What I want to do in this chapter is to look at a number of unresolved tensions in Darwin's own thought. Given the importance of Darwin as a thinker, and his influence even now, this would be worthwhile in its own right; but some of the tensions in Darwin's thought continue to pervade evolutionary thinking to-day. I will not be able to document this here in any detail, and will leave it to readers to draw their own conclusions (though they could consult my *Beyond Evolution*). To avoid misunderstandings at this point, though, I should emphasise that what I say here is not intended to throw into question the concept of biological evolution, but only to point out some difficulties which may arise in a particular understanding of that concept, particularly (as will become evident) when it is applied to human history (as Darwin intended).

1 Evolution and Progress

According to one strand of Darwinian thinking, evolution is fundamentally a relative notion, and there is in Darwin's theory no necessity that the development of evolutionary processes should be progressive in any sense other than "better at surviving and reproducing." Proponents of this interpretation will point out that in *The Origin of Species* Darwin hardly uses the term "evolution" (which definitely has connotations of progress in a more general sense), but tends to speak of the laws of variation, natural selection and descent with modification. "Descent with modification" carries with it no implication that the modifications brought about through natural selection will necessarily be bigger or more complex or more beautiful or more intelligent.

On this austere understanding of what is going on, success in the struggle for survival is all that really counts, and all that natural selection guarantees, and that may come in all sorts of ways. What the theory says is that variations which benefit their possessors in the struggle for survival will do better and eventually displace their competitors and their less successful con-specifics.

But success is always relative to a given environment, and may not require greater complexity or perfection viewed in absolute terms. Thus a longer neck might benefit its possessor if there are tall food bearing trees, but not if the trees all die out. If that happened the very same characteristic that was once an advantage will later prove a disadvantage. This effect can be quite radical in leading to the shedding of costly characteristics within a species when they are no longer required. Thus we see cave dwelling descendents of sighted creatures with no sight, or flightless birds in New Zealand (before humans arrived with their rodent followers). In each case the effort and energy needed to produce sight and flight was not necessary for survival, so the faculties in question simply dropped off, for they had come to constitute a cost with no consequent advantage.

Darwin was well aware of all of this:

> As natural selection acts by competition, it adapts the inhabitants of each country only in relation to the degree of perfection of their associates [...] Nor ought we to marvel if all the contrivances of nature be not, as far as we can judge, absolutely perfect; and if some of them be abhorrent to our idea of fitness.

And having mentioned bees being killed by their own stings, drones being produced in vast numbers for just one act, then to be slaughtered, *ichneumonidae* feeding in the bodies of live caterpillars, and other examples of waste, profligacy and worse in nature, he concludes "the wonder indeed is, on the theory of natural selection, that more cases of the want of absolute perfection have not been observed" (Darwin 1982, 445).

Logically what Darwin says here is impeccable. Relative fitness and non-progressive development, fit enough just for the relevant environment, is all that is strictly implied by the theory of natural selection. Darwin is also keen on occasion to point out that our own ideas of what constitutes perfection in a species might just be a little, shall we say, anthropocentric: he wrote in a letter that while to us intelligence may seem the chief mark of progress, to a bee it would no doubt be something else. This last sentiment might well seem to some to put Darwin in a favourable light, as immune to the race and species progressivism characteristic of his age. Unfortunately (perhaps) Darwin turns out to have had no such immunity, nor did he see evolution in general in strictly relative terms.

This is actually perfectly evident from the closing pages of *The Origin of Species*. "As natural selection works solely by and for the good of each being, all corporeal and mental endowments will tend to progress towards perfection," he writes at the end of the penultimate paragraph. What he says there is something of a *non sequitur*, especially given our earlier observations on the logic of natural selection, which would license no such perfectionist optimism. One wonders, moreover, what Darwins's own standard of progress and perfection is. Is a horse more perfect than a dinosaur, a fish than an amoeba? Is mankind more perfect than the bee? If we think we know what Darwin's answers might be to at least some of these questions, there is more than a hint that in his judgements he would be implicitly judging the animal kingdom by the human characteristics of intelligence, rationality, morality, brain complexity and the rest.

2 Darwinism and the Creator

At the start of the paragraph we have just quoted, Darwin had spoken (as he always did in all editions of *The Origin of Species*) of his system as being in accordance with "the laws impressed on matter by the Creator." We can argue about just what Darwin meant at the various stages of his life by "the Creator"; but it would be hard to have a mind-set which could make any reference, however metaphorical, to a creatorial mind which did not take some tendency towards the better as being inherent in creation. Even granted Darwin's steady drift towards personal agnosticism, in his core theorising there remain significant traces of (dare we say?) design thinking.

It is not just that nature *mimics* human livestock breeders, which is what Darwin argues in his less exuberant moments. In a striking, but not a-typical passage from the Natural Selection chapter in *The Origin of Species*, Darwin says:

> Natural selection is daily and hourly scrutinising, throughout the world, every variation, even the slightest; rejecting that which is bad, preserving and adding up all that is good; silently and insensibly working, whenever and wherever opportunity offers, at the improvement of each organic being... (Darwin 1982, 133)

Natural selection *scutinising, rejecting, preserving, adding, silently working at the improvement of each (!) organic being, and doing it all daily and hourly.* Metaphor, all metaphor, we will be told, no doubt correctly. But metaphors reveal and metaphors are powerful; and this one is all of a part of Darwin's attempt to hold on to natural selection as a progressive, beneficent force, an attempt which all but forces him to envisage it anthropomorphically, as a displaced intelligent designer, doing the Creator's work for Him, through the laws He has impressed on it.

In the closing passage of *The Origin of Species* Darwin says this:

> Thus, from the war of nature, from famine and death, the most exalted object which we are capable of conceiving, namely, the production of the higher animals, directly follows. There is a grandeur in this view of life, with its several powers, having been originally breathed into a few forms or into one; and that, whilst this planet has gone on cycling according to the fixed law of gravity, from so simple a beginning endless forms most beautiful and most wonderful have been, and are being, evolved.

On the other hand, we know that even as early as 1856 Darwin lamented the "clumsy, wasteful, blundering, low, and horribly cruel works of nature." In 1865 he reflected on the certainty of the extinction of all life:

> to think of millions of years, with every continent swarming with good and enlightened men, all ending in this, and with probably no fresh start until this our planetary system has been again converted into a red-hot gas. *Sic transit gloria mundi* with a vengeance.

And in 1881 he told Wallace that even with everything to make him happy and contented, "life has become very wearisome to me," partly surely because of his growing agnosticism (see Greene 1999, 53-54.) For all Darwin's pointing up of aspects of sympathy among us and other creatures, and his talk of grandeur in his vision notwithstanding, one can easily become depressed, as Darwin seemed to be himself, with the tension between these positive features of evolution and his fundamentally cruel and bleak aspect re-imagining of nature.

3 Our Knowledge of Reality

Actually rather more hangs on agnosticism at this point than Darwin's personal mood, as Darwin himself recognised. If natural selection is all that there is, and if the human mind can be explained in purely evolutionary terms, as deriving from that of the lower animals, why should we accept that what we think about ultimate reality has any objective validity? "A dog might as well speculate on the mind of Newton," Darwin wrote in 1881. He went on to express a "horrid doubt" as to whether "the convictions of man's mind, which has been developed from the mind of the lower animals, are of any value or at all trustworthy. Would anyone trust the convictions of a monkey's mind, if there are any convictions in such a mind?" (Letter to William Graham, July 3rd, 1881, in F. Darwin 1888, Vol. I, 285.)

Part of the point here is that just like a bee, a monkey might have a very different perspective on the world from us; and, in the case of scientific and philosophical speculation, compared to us, a very limited one. But equally, ours might seem even more limited to our distant descendents or to creatures with higher intellectual powers. Darwin hopes that natural selection will eventually produce people who would look on him and Lyell and Newton as "mere barbarians;" but while that does seem to follow from the point about natural selection's programme of relentless scrutinising and improving, what confidence would that leave us in the theories of Darwin, Lyell and Newton? Will their theories, in the future, seem no more reliable than those of the primitive and barbaric Tierra del Fuegans Darwin encountered on his epic voyage did to him, and who caused him to remark in his journal for 17th December 1832 on how wide was the difference between savage and civilised man, greater than that between wild and domesticated animals, "inasmuch as in man there is a greater power of improvement"? So what would an improved Darwin or Lyell of 500 years hence think of what will no doubt seem to be the primitive ramblings and superstitions of their predecessors from the 19th century?

If these were not sufficient grounds for scepticism on our existential and metaphysical convictions, we also have to consider the nature of Darwinian explanations. As we have already pointed out, the theory of natural selection tells us that a creature's physical and mental development is conditioned by what will aid survival and reproduction – and that is all. Why are we to suppose that speculating on our own nature has anything to do with that, or, even more, that the faculties we have developed to help us get round the savannah and find mates in earlier times are going to help us in coming to the truth in advanced scientific and philosophical investigations? Or indeed in the looking long into the past and into the future, which Darwin himself saw as part of our capacities? How did these come about through natural selection alone?

Darwin's point is put with telling directness by Thomas Nagel:

> If, *per impossibile,* we came to believe that our capacity for objective theory were the product of natural selection, that would warrant serious scepticism about its results beyond a very limited and familiar range. (Nagel 1986, 78)

Nagel concludes that the development of the human intellect—which can go beyond the limited and the familiar—probably provides a counter-example to the view that natural selection explains everything. I would concur, adding two further points. The first is that even if we add the role of the intellect in sexual selection, saying that our minds have developed partly in order to attract mates through storytelling and other mental performances neither validates those performances or explains why it is that potential mates value those who pursue objective theory (if they do). The theory of natural selection needs supplementing at both these points to give a satisfactory account of our pursuance of objective theory.

Then secondly, as Darwin himself acknowledged, the theory of natural selection is in danger of self-destructing. If that theory explains what we think and do in terms of the value things have for us in promoting survival and reproduction, saying in effect that we accept them because they promote survival and reproduction, the same must be true of the theory of natural selection itself. We accept it, if we do, because it helps us in the struggle for existence, not because it is true, which would of course provide no rational argument against the creationist or the Islamist who might, not unconvincingly, find great support for survival in the following of his creed. So Darwinism undermines its own claims to be true (just as in analogous ways do the theories of Marx, Nietzsche and Freud if we take them at face value).

4 Savages and Civilized Races

Let us, though, for the moment assume that Darwinian explanations are in general true and do not apply to the Darwinian theory itself, and let us see what that theory implies about human development. It turns out that Darwin's early reaction to the Tierra del Fuegans was not an anomaly, but is all of a piece with the core doctrine of continuous improvement through natural selection. He refers to them again at the end of *The Descent of Man* in 1871:

> They possessed hardly any arts, and like wild animals lived on what they could catch; they had no government, and were merciless to everyone not of their tribe.

And opening out his discussion, he goes on to say that he would prefer to be descended from a monkey or a baboon who manifested traits of loyalty and self-sacrifice as from

> a savage who delights to torture his enemies, offers up bloody sacrifices, practices infanticide without remorse, treats his wives like slaves, knows no decency, and is haunted by the grossest superstitions. (Darwin 1898, vol. II, 440)

Earlier in the main body of *The Descent of Man* Darwin had written a whole chapter on the way inferior races had been replaced by superior ones; even "at the present day civilised nations are everywhere supplanting barbarous nations, excepting where the climate opposes a deadly barrier" (Darwin 1898, vol. I, 197). Indeed part of Chapter V of *The Descent of Man* is devoted to rebutting the contrary suggestion that all races started at the same level, with some declining over time. For Darwin, as an evolutionarily progressive thinker, the descent of man implies *ascent* both from lower species and from lower stages of human development. As early as October 11th 1859, in a letter to Chrarles Lyell, Darwin had written "I look at this process as now going on with the races of man; the less intellectual races being exterminated" (Darwin 1983-2004, Vol. 7, 345).

As late as 1881 Darwin wrote:

> The more civilized so-called Caucasian races have beaten the Turkish hollow in the struggle for existence. Looking to the world at no very distant date, what an endless number of the lower races will have been eliminated by the higher civilized races throughout the world.

Admittedly this is in a letter (the same one, in fact, already referred to), but that remark (redolent as it may be of the contemporaneous talk of "Bulgarian atrocities" and of "sick men of Europe") is precisely in reply to a correspondent who was doubtful that the struggle for survival and natural selection had done much to contribute to human progress. In 2010 it is hard not be disturbed Darwin's casual reference to the elimination of endless numbers of lower races, and even more by the way this sort of thinking was taken up by his followers such as Haeckel and von Treitschke, who in turn influenced Hitler. Moreover the remark in question is all of a piece with the teaching of *The Descent of Man*, even if more forcefully expressed.

The question we have to face here is not whether Darwin held the views ascribed to him. He clearly did. The question is whether those views follow from the theory of natural selection. The inescapable conclusion is that, if that theory is to be applied to human history, it is hard to see how, in some form, it can fail to do so.

The very first chapter of that book is entitled "The Descent of Man from Some Lower Form," so clearly no species egalitarianism there. It is indeed just what we had been led to expect from the conclusion of *The Origin of Species* where we were promised that the theory of natural selection would through light on origin of man *and his history*. For if natural selection is a doctrine of progress and if it applies to human history as well as to human origins we must expect that humans will be better than animals in significant respects and that some humans will be significantly better than others. There is, of course, an elision in Darwin's thinking between better *in the struggle for existence* and more civilized. Bulgarian atrocities aside, there is no necessity here. Indeed, after the somewhat chequered history of the twentieth century (making "Bulgarian atrocities" seem rather small beer), it should be more clear to us than it may have been to Darwin is that in human affairs the struggle for existence is not in itself a guarantee of progress in any dimensions other than those of surviving and reproducing.

5 Darwinism and Social Policy

More clear, but Darwin was not entirely unworried in this area. For he did, like many of his contemporaries, notice a tendency in his time for the unfit, the inferior "in body or mind" and even the abject poor to breed, and, though he does not say this explicitly, possibly to outbreed the prudent and the strong. If mankind is to advance, we must uncover the laws of inheritance and then legislate against marriages among the inferior. We must encourage the poor not to marry (for abject poverty "tends to its own increase by leading to recklessness in marriage"), while at the same time urging the prudent and the most able to rear the largest number of offspring. Above all we must ensure that the struggle for existence is not softened in its severity by laws and customs: "otherwise (mankind) would sink into indolence, and the more gifted men would not be more successful in the battle of life than the less gifted" (Darwin 1898, Vol. II, 438–440).

There is in fact a degree of tension in Darwin's own mind at this point, because as well as the struggle for existence, he wants our moral qualities to be developed (partly because he believes that a group with a strong communal morality based on mutual sympathy will do better than less coherent groups). But might it not be just those moral qualities which recognise a common good, which protect the inferior and the poor against the most severe effects of the struggle for existence, which might then undermine human progress (on his view)? Indeed it is just so. "It is surprising," Darwin observed, "how soon a want of care [...] leads to the degeneration of a domestic race; but excepting in the case of man himself, hardly anyone is so ignorant as to allow his worst animals to breed." But in our own case, and for moral reasons, a degree of such "ignorance" must be tolerated. We must, Darwin admits, "therefore bear the undoubtedly bad effects of the weak surviving and propagating their kind," while at the same time doing what we can to ensure that "the weaker and inferior members of society do not marry so freely as the sound" (Darwin 1898, Vol. I, 206). So the general import of his message is clear. We must take as much care in our own marriages as we take in the breeding and selection of our domestic animals, and we must also maintain social structures which allow untrammelled competition; both these injunctions follow pretty directly from applying the theory of natural selection to human society, as does the view that societies are to be ranked in degrees of success.

6 Conclusion: An Anthropic Approach to Evolution

We have been considering a number of aspects of Darwin's theory. In so doing we have found a tension between the theory of natural selection taken strictly and things which Darwin clearly holds strongly and wants to say. In particular we have found difficulties with his view of evolutionary progress, with his view of our own mental capacities, with his attempts to rank human societies, and also with what he considers desirable within human societies. To put it bluntly, natural selection gives no warrant for any progressivism regarding evolution. It makes it hard to see what faith we should have in our scientific and philosophical speculations. It gives no warrant for associating success in evolutionary terms with a greater degree of civilisation. At the same time the theory of natural selection seems to sanction a type of society which would run counter to many commonly held moral virtues and decencies. The interesting thing is that in each case Darwin himself gives sign of straining against the strict view of natural selection, and of wanting to promote a less austere view of things.

In what we have come to see as the austere version of evolution, that delimited by Darwin's strict theory of natural selection, the picture which is given is of life being a desperate struggle by individuals to survive in an environment which if not actually hostile is largely indifferent to them. The key levers in this drama are random variations within the individuals and selective retention of a few of them by an environment which cares nothing for any of it. We are obviously a long distance from Darwin's own sense of natural selection carefully scrutinising, selecting, preserving, ceaselessly and silently working for the good of all and each, and it is difficult to see where any such notion could gain a foothold. Nor is there any sense that the process as a whole is likely to move in a progressive direction, towards greater intelligence, complexity and morality. Rather to the contrary, the universe looks far more like that described by Jacques Monod in *Chance and Necessity*: "The universe was not pregnant with life, nor the biosphere with man" (Monod 1971, 145). We human beings are here by chance, in a universe which is not responsive to us at all, and within which our existence has no significance. For Monod, mankind is a gypsy, living in an alien world, which is deaf to his music. (He apparently saw no difficulty in having this alien world giving rise to creatures (us) who are able to conceive the world and their activity in terms of values.)

In recent years, as is well known, the view that the universe was not pregnant with life and consciousness has been challenged by what has become known as the anthropic principle. It is obviously true, tautologically so, that, given that we are here, the universe must be such and must have been such as to allow for the existence of intelligent knowers, such as ourselves. It turns out, though, non-tautologically, that a very high degree of fine tuning even at the start of the universe, would have to have been in place in order for intelligent life (us) to have been possible. Can anything be concluded from the fine tuning point?

At the very least, it suggests that Monod's basic stance needs qualification. From the very beginning, the universe was, if not pregnant with life, certainly ready for the emergence of life. And the more precise the fine tuning and the more etched into the substance of things that fine tuning is, given the immense amounts of time and space involved for things to work themselves out, the closer readiness becomes to pregnancy. In a universe of the extent of ours, it is not unrealistic to think that possibilities embedded in the universe's basic structure are highly likely actually to occur. It is reasoning of this sort which leads the adherents of the so-called "strong" anthropic principle to conclude that life and mind do not have to be imported into the universe from outside or by chance. They are "etched deeply into the fabric of the cosmos, perhaps through a shadowy half-glimpsed life principle" (Davies 2006, 302-303). Given the notorious problems in explaining life and consciousness in purely physicalistic terms, such a view is not just helpful in general terms. The difficulties themselves might actually open us to the possibility that some such thing must be true, that life and mind are there, embryonically, right from the start—otherwise it becomes well-nigh impossible to see how they could have arisen.

We do not have to acquiesce in the more colourful conclusions drawn by advocates of anthropic thinking to see its basic orientation as suggestive in a number of ways. If the universe is disposed to produce life and mind right from the start, we will no longer see ourselves as tangential to it, a mere random accident in a fundamentally lifeless system, gaining whatever knowledge we have of it as a chance side effect of our striving to survive in it. If our mental faculties are rooted in the fabric of the universe, it will not be surprising or problematic if they do deliver knowledge of it way beyond the basics we need for survival. If the universe as a whole is evolving forms of life and mind, the progressive thrust of evolution in that direction will not be such a mystery. Also, if life and mind are themselves goods from the point of view of that evolution, we may well be led to value states of feeling and mind for their own sakes, and not simply as aids to survival and reproduction. Indeed something more than survival and reproduction may come to be seen as implicit in nature from the start; so our own tendencies to morality and co-operation (genuine altruism) will no longer seem the anomaly they will inevitably are if nature is conceived in strictly Darwinian terms. Finally an approach to evolution and life which stresses complexity and mutual belonging will be far less concerned to emphasise struggle in the way Darwin does, which will obviously have ecological and ethical resonances.

It is no part of the argument of this chapter to suggest that what we have been calling the austere version of Darwinian thinking cannot be defended scientifically. But what I would suggest is that it is not scientifically demanded either. The facts adduced by Darwin to support his theory do not point unambiguously in either direction, which is part of my point in underlining the tensions which exist within Darwin's own world-view, and which are revealed in his own words. To put this point in terms familiar to contemporary philosophy of science, here, as elsewhere, theory is under-determined by data; which suggests that factors beyond the data have a role in determining the way we (and Darwin) might chose to read our data.

References

Darwin, C. (1898). *The Descent of Man*. London: John Murray.
_____ (1982). *The Origin of Species*. Harmondsworth: Penguin Books.
_____ (1983-2004). *The Correspondence of Charles Darwin*. Cambridge: Cambridge University Press.
Darwin, F. (ed.) (1888). *The Life and Letters of Charles Darwin*. London: John Murray.
Davies, P. (2006). *The Goldilocks Enigma: Why Is the Universe Just Right for Life?* London: Penguin Books.
Greene, G. C. (1999). *Debating Darwin: Adventures of a Scholar*. Claremont: Regina Books.

Monod, J. (1971). *Chance and Necessity.* New York: Knopf.
Nagel, T. (1986). *The View from Nowhere.* Oxford: Oxford University Press.
O'Hear, A. (1997). *Beyond Evolution: Human Nature and the Limits of Evolutionary Explanation.* Oxford: Oxford University Press.

3

Evolution, Cognition and Value: the Ingredients for a Naturalist Philosophy

Konrad Talmont-Kaminski

1 O'Hear's Tensions

This chapter is written in response to Anthony O'Hear's "Darwinian Tensions." However, it goes beyond a mere response in attempting to present some of the main aspects of a robust naturalism which finds its basis in evolutionary theory. Central to it is a naturalist account of cognition based upon an investigation of the deep analogies between cognition and evolution.

In many ways O'Hear's chapter provides a fine foil for such an account. O'Hear manages to touch on many of the issues that help to differentiate naturalised philosophy from its antinaturalist predecessors. That he does so in the context of Darwin's theory of natural selection is particularly useful given that evolutionary theory has come to play a central role within naturalised philosophy. For these reasons, my chapter follows the structure of O'Hear's text. Thus, in the first section I will discuss the question of whether the notion of universal progress is a part of the evolutionary theory, showing that O'Hear's assertion to the affirmative is incorrect. In the second section, I consider the use of anthropomorphic language in talking about evolution and show how it is best explained as an effect of the human mind having evolved to deal with the social environment, rather than being explainable in terms of Darwin writing under the influence of a pseudo-theistic atavism. The third section is the longest, as in it I deal with the question of how evolutionary theory can explain the human ability to use our reason to deal with profoundly novel situations. This requires that I pursue the similarities between evolutionary and cognitive processes; both being open-ended yet, at all times, bounded. The fourth section turns to a discussion of the ethical implications of evolutionary theory, and argues that to think that there are any is to commit the naturalistic fallacy. As I show in the final, fifth, section, the strong anthropic principle provides no alternative to the view presented herein, leaving us very much in the position that O'Hear finds clearly daunting but which to others, including myself, is bracing.

Unfortunately, at the same time as O'Hear's chapter is a useful foil it is plagued with just the kinds of tensions that he unsuccessfully searches for within Darwin.

Firstly, O'Hear states that he is merely aiming to point out some unresolved tensions in Darwin's writings and that he will not discuss their significance to modern evolutionary thought. This makes it sound like he is about to present an essay in the history of science. If so, the fact that his article concludes with an argument for taking seriously the strong anthropic principle—a conclusion he makes no reference to in his introduction—is quite surprising. Are we to understand that he thinks this principle a viable alternative to the views presented by Darwin over one hundred and fifty years ago or that it is a viable alternative to modern evolutionary theory? If the first is true, hardly any modern scientist would be concerned, even if O'Hear's argument were sound. Though Darwin is greatly respected, just as Newton or even Archimedes, science does not move forward by defending the views of great scientists but by improving upon them. For this reason, biologists do not read *The Origin of the Species* in the same way that philosophers may read *The Critique of Pure Reason*—it is a historical text rather than one to have an ongoing debate with. So, a biologist could accept this reading of O'Hear's argument and simply respond that it shows another problem with Darwin's original formulation of the theory of evolution. However, what if, by "leaving it up to the readers to draw their own conclusions" O'Hear means to claim that the strong anthropic principle provides something of an alternative to modern evolutionary theory, as many of his statements suggest. In that case, his argument seems, at best, enthymematic at the very point it needs to be most explicit. To make it more than that it would be necessary for O'Hear to show that the aspects of Darwin's thinking that O'Hear discusses are essential to modern evolutionary theory rather than, for example, mistaken representations of it in popular culture. One basic way in which O'Hear might attempt to do this is by showing that biologists do seem to exhibit a particular allegiance to Darwin. To do that, however, would be to misunderstand the nature of this allegiance in that it is based upon a respect for Darwin's achievements and a feeling of shared goals rather than a faithful adherence to Darwin's views. As such, if O'Hear's aim was to say something of relevance to modern evolutionary theory he should have aimed his arguments at its real, contemporary, empirical basis. What might have been an appropriate critique of a philosophical tradition is seriously wrongheaded when aimed against a scientific theory. Unfortunately, in so far as O'Hear appears to read Darwin as a philosopher rather than a biologist, he does appear to be falling into the genetic fallacy.

Secondly, O'Hear is very clear in denying that he is rejecting biological evolution and claims to be merely pointing out some shortcomings of one way of understanding evolution. It is as well that O'Hear does this as many of the arguments he presents are stable fare for creationists. Drawing attention to this fact comes close to an ad hominem response, except that it is important to recognise that, due to the notoriety of the creationists, these arguments have been looked at in detail by many scientists and philosophers. Unfortunately, O'Hear makes no effort to take into account or even acknowledge those criticisms (SkepticWiki.org provides a useful introduction to these). This is particularly problematic given that the criticisms are compelling, to say the least. Although O'Hear never does present his alternative understanding of evolution, present throughout the paper is the suggestion that a more "positive, optimistic" alternative is in the offing. Given what O'Hear writes, it seems reasonable to conclude that the alternative he has in mind is what is known as theistic or guided evolution—the view that evolution has taken place but that, instead of being a completely natural process, it has been interfered with by God. It is arguable whether this is still biological evolution, given its necessary supernatural baggage. More importantly, however, similarly to creationist positions, this view has been thoroughly critiqued by biologists and been found lacking in the empirical evidence that is essential for any scientific claim, as well as having been critiqued by philosophers and been found to offer no real explanation (see Dawkins 1986, Dennett 1995).

O'Hear's desire for an "optimistic" alternative leads to the final, and most troubling, tension in his work. He is very critical of the belief in universal progress that he perceives in Darwin's account of evolution. Yet, he himself would wish to believe that the universe was fine tuned to make possible intelligent life. If that were the case, the whole history of the universe would be that of the progress toward this goal and the resultant notion of evolution would be profoundly progressivist. If that is so, however, O'Hear ends up arguing for a view of evolution that has the worst problems that he thinks Darwin is weighed down by, including "race and species progressivism."

It is by considering whether Darwinian evolution really does require progressivism that we will begin.

2 Progress

O'Hear's first concern is to establish that Darwin worked with an absolute notion of progress. To show this, he quotes from *The Origin of Species* where Darwin writes that "As natural selection works solely by and for the good of each being, all corporeal and mental endowments will tend to progress towards perfection." This leads O'Hear to ask whether, according to Darwin, a fish is more perfect than an amoeba, and humanity more perfect than bees. Darwin dealt with that question in Notebook B, however, where he wrote that it is "absurd to talk of one animal being higher than another." Not surprisingly, then, a fairly quick investigation of Darwin's use of the term "perfection" within *The Origin of Species* reveals, that it is consistently used to describe features of an animal, where such features are deemed to be perfect on the basis of how well they serve that animal in its environment. A fine example of this is Darwin's discussion of the perfection of the eye in chapter six of *The Origin of Species*. In other words, the notion of perfection that Darwin is working with is one that is already relativised to a given animal and its environment, as is actually clear in the quote used by O'Hear. Talk of one species as superior in some absolute sense to another is not something that Darwin engages in. This is highly significant as it shows that O'Hear's argument falters at its first step and that Darwin's use of the notion of perfection does not commit him (or natural selection) to "the race and species progressivism characteristic of his age" that O'Hear accuses him of. Indeed, any such progressivism appears to be quite foreign to it.

Given that O'Hear leaves it up to the reader to decide what significance Darwin's views have had for modern evolutionary theory it is important to note that, in this respect, evolutionists have on the whole continued to reject progressivism. The relative notion of progress that Darwin worked with is the one that evolutionary biology has continued to rely upon, with any attempts to argue for evolution being progressive on a grand scale belonging to the sidelines of scientific evolutionary debates (even though such notions have plagued popular understanding of Darwin's work, as well as its early misapplications in social sciences). If anything, biologists have moved even further away from any absolutist notions in so far as talk of perfection, even in a relativised form, has been replaced by talk of "well adapted" traits. Indeed, the eye, considered by Darwin for its "perfection," is now the standard example for showing how even very complex and highly functional organs are jerry-built end products of variation, selection and retention, riddled with slapdash solutions that no competent engineer would be willing to put their name to.

Richard Lewontin (1982, 165), one of the foremost theoretical biologists of our times, states the current view unequivocally:

> Evolutionary theory in general no longer incorporates notions of progress or of unidirectional change. Evolution, at least in the modern view, is going nowhere in particular. Older ideas that evolution has led and is leading to greater complexity, greater homeostasis, greater stability, are now seen as vestiges of a 19th century progressivism that is without empirical foundation in the history of life.

3 Creator

Having shown to his own satisfaction that Darwin was a progressivist, O'Hear goes on to discuss Darwin's use of anthropomorphic language to describe the workings of natural selection. O'Hear's thesis is that for Darwin natural selection played the role of a surrogate Creator, ensuring that evolution would lead to progress. O'Hear claims that such a "positive" view of evolution is in tension with Darwin's view of nature as red in tooth and claw. Of course, even if O'Hear's claims of progressivism were true, this conclusion would not follow. To the degree that evolution is progressive (locally or globally), any progress is due to the cruel winnowing away of those organisms that are less fit—it is thanks to selection that evolution is not a random process, contrary to what some of its opponents claim. Darwin's analogy between the workings of evolution and human breeders is most apt in this context in that a breeder obtains pigeons with striking new plumage by selectively allowing the most promising birds to procreate while he breaks the necks of the rest. One does not need a Panglossian view of the world to appreciate its beauty.

It may be worthwhile to consider at this point an example of natural selection at work (see Dawkins 1986, 178ff for the original discussion of this example). Evolutionary arms races are a common phenomenon in which changes in a particular trait of one organism alter the environment of another organism in such a way as to put selective pressure on a trait of that organism, with the changes in the second organism feeding back to produce selective pressure that further changes the original trait of the first organism. One simple example of this phenomenon is the arms race between the cheetah and the gazelle. In both cases, the trait that is strongly selected for is increased speed, the result being that the cheetah is the fastest land animal. The graceful sprint of the cheetah has been made possible, however, by countless generations of predators that starved for the wont of a gazelle they could catch. The speed of the cheetah might be considered some kind of perfection, except that the cheetah and the gazelle are both only adapted to the particular environment that includes both species. Due to the enormous metabolic cost borne by both species in developing and maintaining their capacity for speed, they would be quickly outselected by other species were their main prey or predator to disappear, making speed much less adaptive—an eventuality that has largely come to pass due to the actions of homo sapiens. In this example we can easily see the lack of a tension between progress and the manner in which it arises as well as that any progress is quite local.

While this shows there is no tension between the horror and the beauty of the natural world as understood by Darwin, it does not deal with O'Hear's claims concerning Darwin's use of anthropomorphic language and references to the Creator. To understand that, it is necessary to consider the historical and evolutionary context of Darwin's work. Darwin was well aware that most of his readers would be largely ignorant of biology and culturally steeped in creationist myths. He realised that this would make it very difficult for them to accept evolution on the strength of its evidence. His fears have been borne out by the fact that even now, a hundred and fifty years after the publication of *The Origin of Species* and long after evolutionary theory has become the foundational theory for all life sciences, it continues to be opposed by fundamentalist Christians in the US, as well as by other groups around the world whose opposition is invariably religiously motivated. In effect, Darwin knew that he would have to speak to the theists who would see in evolution a threat. Thus, in the final section of *The Origin of Species*, from which O'Hear's quote comes, Darwin actually claims that evolutionary theory "accords better with what we know of the laws impressed on matter by the Creator" than the idea that each species was independently created. His reference to a deity is, therefore, aimed at convincing theists that there is no conflict between theism and the view that humans are the product of fully natural evolutionary processes—a position that has come to be accepted by many theists.

Why does Darwin use anthropomorphic language in speaking about natural selection, in general, however? The thing to note is that anthropomorphic language is hardly a rarity among humans. We commonly treat inanimate objects as having humanlike mental properties or as exhibiting humanlike behaviour. We get angry with our cars when they ignore our entreaties to start on a cold morning. We speak of the sea as angry and cruel. The human predilection to anthropomorphise does not stem from a need to seek out deities in such objects. Indeed, the connection runs in exactly the opposite direction. The human mind has evolved the capacity to deal with complex social situations. The tools it uses to do this are available to deal with other situations as well. Very often, being able to express a situation that involves no agents in terms that do helps us to be able to understand it, as doing so gives us immediate access to a plethora of social intuitions that direct us to the correct conclusion so long as we have appropriately conceptualised the situation. This is precisely the property of the human mind that Darwin is making use of when he describes natural selection in terms of intentional actions. Of course, even though useful, this approach is dangerous and can go awry. Recent research in cognitive science of religion suggests that human beliefs in supernatural agents are partly due to a tendency to look for agents in the most unlikely places (Barrett 2000 is a useful introduction). Such a hyperactive agent detection device, as it is called, will have been very useful in detecting hidden predators but also produced numerous false alarms. Some of these led to beliefs in forest spirits and, through a complex set of evolutionary and cultural processes that are being teased apart at this very moment, to acceptance of the idea that there exists a Creator. In short, it is not that natural selection is Darwin's God-surrogate but that belief in a Creator is a by-product of the evolution of our natural cognitive abilities. This is a good example of how evolutionary theory is not just capable of dealing with objections to it but can even tell us something about the reasons why such objections may seem convincing to our, evolved, minds.

4 Cognition

The frailty and fallibility of human cognition research reveals may well lead one to conclude that O'Hear is quite right when he asks:

> If natural selection is all that there is, and if the human mind can be explained in purely evolutionary terms, as deriving from that of the lower animals, why should we accept that what we think about ultimate reality has any objective validity?

There is much to be said for this worry. Certainly, for example, we ought to be very wary of our intuitions (for a discussion, see Bishop & Trout 2004, Feltz & Bishop 2010). Any general scepticism about evolved human reason is misplaced, however.

O'Hear wonders whether in the future Darwin's theory will "seem no more reliable than those of the primitive and barbaric Tierra del Fuegans Darwin encountered on his epic voyage seemed to him." Before exploring this worry let us ask ourselves just how "unreliable" the beliefs of "primitive" and "barbaric" peoples are. Jared Diamond (1997, 143) provides something of an overview:

> An entire field of science, termed ethnobiology, studies people's knowledge of the wild plants and animals in their environment. Such studies have concentrated especially on the world's few surviving hunting-gathering peoples, and on farming peoples who still depend heavily on wild foods and natural products. The studies generally show that such peoples are walking encyclopaedias of natural history, with individual names (in their local language) for as many as a thousand or more plant and animal species, and with detailed knowledge of those species' biological characteristics, distribution, and potential uses.

Diamond admits, a hunter-gatherer thrown into the middle of a modern city will be at a loss and seem foolish but this speaks not of the unreliability of their knowledge but of it being limited to their natural environment. A city-dweller is as much at a loss in the wilderness. In 1860 the Burke and Wills expedition set out to cross Australia from Melbourne on the south coast to the Gulf of Carpentaria on the north. Only one man managed to survive the whole trip even though the land the expedition travelled through was inhabited by native Australians that actively assisted the travellers until they were fired upon by one of the Europeans. If anything, Diamond argues (1997, 20–21), the hunter-gatherer may be smarter than the city-dweller due to ongoing selection pressures. Still, both are members of the same species. So, perhaps, it is hardly surprising that their knowledge of their environments is of similar reliability. A very different picture might be expected to be revealed in the case of the "lower" animals.

In 1966 MacArthur and Pianka put forward optimal foraging theory, which aimed to analyse animal foraging behaviour in terms of how it compares to optimal strategies. It has been found that once environmental and behavioural constraints are taken into account, even very simple organisms are able to approximate optimal strategies. This is not to suggest that such animals possess the powers of reflective cognition but that their knowledge of their environment is reliable though, once again, limited. Indeed, animals' capacity to obtain information about their environment is so reliable that even in today's world of advanced technology the airport customs officer looks for drugs accompanied not by a mechanical detector but by a beagle. Studies of animal behaviour are full of examples of the inventive ways in which various non-human animals make highly efficient and inventive use of their environment and the information available to them. Of course, just as with the humans, other animals that find themselves in significantly altered environments will end up acting in ways that are systematically disadvantageous. Thus, for example, baby tortoises born on the beaches near housing developments often crawl up the beach toward the lights of the houses, instead of crawling down toward the moonlight reflected in the water they, ultimately, seek.

By analogy to these examples, we can see that, if Darwin's theory will come to be looked upon as "primitive" in the future, this will probably not mean that it will be rejected but that it will be found to be a mostly accurate description of a limited range of phenomena, just as Newton's physics has been found to accurately describe relatively slow movements of macroscopic objects, failing as speeds approach the speed of light. This has not, by any means, revealed Newton to be a "mere barbarian." Even as they recognise the shortcomings of his theories, modern physicists respect Newton highly for the partial understanding he was able to achieve. Indeed, given the recognised shortcomings of Darwin's understanding of evolution—he lacked a model of the genetics that underlie heredity—this is something like the attitude modern biologists already have to Darwin.

Everywhere we bother to look, the same pattern appears—the knowledge that is possessed by an organism is quite reliable in its natural environment but fails outside of it. This is as true of human scientists as it is of the paramecium. Yet, clearly, it is possible to partially transcend existing limitations both by evolutionary and by cognitive means. Thus, in the case of evolution, novel species sometimes gain a massive advantage from having access to new sources of information such as a improved sensory organs or improved ways of utilising this information in their interactions with their environment. For example, it has been suggested that the cause of the Cambrian explosion was the original evolution of the eye, leading to rapid speciation within the clade or clades that had developed eyes (Parker 2003). Similarly, humans and other animals are capable of transcending limitations to varying degrees. Thus, finding themselves in a novel environment, humans (and many other animals) are often capable of changing their behaviour to better suit that environment, *given sufficient time to learn about it.*

O'Hear remarks on this (according to him purely human) capacity and, echoing Thomas Nagel, concludes that it could not be explained by evolution. Nothing could be further from the truth. Already, we see that evolution, just as cognition, is capable of transcending limitations—a point that is probably sufficient to establish that the conclusion drawn by O'Hear does not seem to follow. Moreover, non-human animals with advanced cognitive abilities, such as chimpanzees or dolphins are also capable of going "beyond the limited and the familiar" though, of course, not to the degree that humans are. The capacity of various non-human animals to do this has been studied at great length by ethologists and others (for an introduction, see Pearce 1997). Indeed, some capacity for learning appears to be present in most animal species, with only very simple life forms (if any) lacking it altogether. On the other hand, in many respects people show an astonishing inability to learn to deal with a new environment. Human desire for fat and sugar formed in an environment in which both were nearly always hard to find. Chronic overeating was not a common threat and a "sweet tooth" was adaptive. In the modern world of convenience shopping, fat and sugar are readily available yet humans have not ceased to be attracted to them, with the result that obesity is the prime cause of lifestyle diseases such as diabetes, heart disease and many cancers.

To understand the basic picture that I am presenting it is necessary to consider what may seem to be two contradictory traits. The first of these is boundedness, exhibited as it is by the cognitive abilities of all animals, including humans. Being bounded, these abilities are only applicable within a limited context, typically the context of the natural environment of the animal that possesses them. Not accidentally, this is precisely the same relativisation as we saw in the case of adaptations that we considered in the first part of this chapter. The reason is that cognitive abilities quite simply are a subset of adaptations and, just like them all, depend crucially upon being used in the appropriate context. The second trait is open-endedness, the capacity to develop in such a way as to transcend existing limitations. An animal such as a squirrel may be quite limited in the degree to which its cognition is open-ended. It may learn basic facts about its environment and adjust its behaviour to suit, avoiding backyards with dogs, going back to dig out nuts it stored for the winter or opportunistically trying out new sources of sustenance. Human cognition is open-ended to a much greater degree, however. We are able not just to learn particular facts but to develop and use novel ways of thinking about the world (de Sousa 2007 explores this property of human cognition while also unambiguously opting for a naturalist explanation). This might seem to be the capacity which, according to O'Hear, goes beyond what is explainable by evolution. That the opposite is true, it is necessary to understand the capacity in the context of a theory, which is sufficiently powerful to talk about both cognition and evolution.

Herbert Simon's bounded rationality theory treats human reasoning as constituted by a set of heuristics that work together to allow us to interact effectively with our environment. As Simon shows in *The Sciences of the Artificial*, the basic notion of a heuristic that he uses is broad enough to apply not just to cognitive mechanisms but to cognitive, behavioural and physiological adaptations in general (Simon 1996). At the same time, it is quite powerful in that it predicts that all heuristics, be they patterns of human reasoning or of the behaviour of simple organisms, will share a number of basic traits. These traits include boundedness (see Wimsatt 2007 for a convenient listing of the traits), the immediate cause of which is that all heuristics necessarily make substantive assumptions about the context in which they are to be used.

The open-endedness afforded by the theory of bounded rationality goes beyond the traits of individual heuristics and consists in the capacity to develop new heuristics (Hooker 1991 develops another account of the open-endedness of cognition). In the case of evolution this is made possible by the mechanisms of variation, selection and retention. With each new paleontological find, our knowledge of how various adaptations arose is becoming ever more complete, in some case already providing us with a veritable slide-show of the intermediate forms that led to supremely functional outcomes. In the case of reasoning, the mechanisms are only now being revealed through work in a variety of the sciences of cognition as well as work in evolutionary explanations of human behaviour (for an introduction to the latter, see Laland & Brown 2002). The basic idea is that, while some heuristics may be in some way hardwired into our brains through evolution (the heuristics investigated by Kahneman and Tversky would probably fall into this group), there is not a set range of heuristics that cognition is limited to relying upon. People are capable of using their limited knowledge of their environment to identify problems with existing heuristics and to develop new heuristics that extend their abilities. Paramount examples of this capacity fill the history of scientific methodology. Thus, the double-blind method that has become standard for testing pharmaceuticals, is the result of successive developments in experimental methods that took place during the twentieth century and were a reaction to the realisation that it is not enough to simply provide a control group which is not subject to the treatment being tested. The initial use of single-blind testing, in which the researchers knew whether they were administering the placebo or the treatment under examination, also turned out to be insufficient in some cases. The reason is that the placebo effect is subtle enough to affect outcomes due to the attitude of the researcher interacting with the subject, even though that person may not be consciously altering their behaviour.

The vital point is that knowledge of the environment plus existing heuristics is sufficient to develop new heuristics. There is no need for a general heuristic-generating mechanism. Such a mechanism would be a "skyhook," to use Dennett's famous term, and as a result would go beyond evolution (Dennett 1995). It seems, therefore, that O'Hear must be assuming some skyhook is necessary to explain human ability to transcend existing cognitive limitations. Unfortunately for that line of thought, such a skyhook is not only unnecessary but, as Dennett shows, ultimately incapable of providing a satisfactory explanation (rather than a merely psychologically satisfying one): It falls foul of the same kind of regress counterarguments that, for example, plague the cosmological argument (Hume 1779). It is, therefore, fortunate that evolution does provide us with the "cranes" sufficient to understand how human reason could have come to exist. Both evolution and cognition exhibit the same traits, the basic reason being that they are both processes of emergence, the only kinds of processes that could have led to the appearance of the kind of complexity that is necessary to support conscious, rational thought (see Bickhard & Terveen 1995).

Going further in his criticism, O'Hear claims that, even if evolution could explain human cognitive capacities, it would cast doubt on our ability to obtain "the truth in advanced scientific and philosophical investigations," including concerning evolution itself. According to him, therefore, evolutionary theory ultimately undermines itself in much the same way that relativist claims that there is no objective truth undermine their own claim to be preferable to more objectivist views of truth. O'Hear's position here has two parts. The first is the question of whether abilities developed to "help get us round the savannah" can be helpful when dealing with quarks or the distinctions between deontology, virtue ethics and consequentialism. This point has already been dealt with when it was argued that both science and philosophy build upon existing human cognitive abilities by developing novel heuristics resulting in a continuum of methods that goes from simple stimulus response to advanced logics (see Haack 2005 for an explicitly Peircean rendering of this continuum). This leaves the second question; the issue of why we should trust a capacity which evolved to allow us to interact in adaptive ways with the environment to provide us with a true description of that environment.

The first thing to be said is that the truth of our beliefs is not normally coincidental to the success of the actions we undertake on their basis. The hunter who comes home with a deer's corpse does so because he was able to locate that deer. Of course, he could have come across it by accident, but anything over random success would provide an evolutionary advantage and would require that beliefs about where deer may be found to some degree reflected the truth. Often it is difficult to express exactly what is true about the beliefs but this problem seems to have more to do with the way analytic philosophy attempts to understand mental states in terms of propositional attitudes than with any underlying question of accuracy. Yet, there are contexts in which it seems plausible that false beliefs are systematically advantageous (for a recent discussion of the evolution of false beliefs see McKay & Dennett 2009 and commentaries, especially Talmont-Kaminski 2009b); indeed, O'Hear explicitly mentions one when he brings up the theist for whom her creed, whatever it happens to be, provides great support. The by-product explanation of religion I mentioned earlier and which is put forward by many researchers working within cognitive science of religion is only one evolutionary account of religion that is being pursued. Another approach is being developed by David Sloan Wilson (2002), who argues that religion is adaptive. Wilson claims that religious beliefs, though false, promote behaviour that is pro-social, leading to stronger communities that are better able to compete against other communities. It seems quite probable that some version of this claim is correct and does not necessarily contradict the by-product account, as the pro-social function of religion is likely to be an exaptation of an existing by-product tendency to treat seriously supernatural claims (Talmont-Kaminski forthcoming).

This might seem to lend credence to O'Hear's worry that evolved reason is unlikely to attain true beliefs. To see why it does not it is necessary to consider some of the very special characteristics of religious beliefs. Unlike most beliefs, including the hunter's belief about deer, the utility of the behaviour motivated by religious beliefs does not depend upon the degree to which they accurately describe the world. It does not matter if people worship Christ, Zeus or Baal, so long as the communal act of worship helps to sustain cohesive faith communities that cooperate on every day of the week, rather than just on Sunday. In so far as some religions lead to more advantageous forms of behaviour, it is not due to their claims being truer but, simply, due to them being better at motivating pro-social behaviour. This divorce of truth and function is made possible by the nature of religious claims. False claims concerning deer mean fewer deer steaks for dinner, a readily observed result. Whether one faces west or east during prayer does not lead to any regular differences in its effectiveness, however. The situation changes, of course, if one's community deems a particular direction to be the correct one. The flip side of this dissociation is the divide between a religious claim and any possible empirical counterevidence. It is no accident that religious beliefs typically concern purported spheres of existence whose supposed effect upon normal life is either hard to discern or so ambiguous as to fit with any outcome. Indeed, successful religions protect their core claims against real world exigencies not just by having largely unfalsifiable content but in two further ways (Talmont-Kaminski 2009a). The first is to make investigation of such claims difficult by deeming them sacred. The second is to oppose the development of scientific methods, which would increase human capacity to investigate the world. Such protection is necessary for two reasons. Firstly, even given how limited human cognitive abilities are, beliefs that fly in the face of experience find it difficult (though not impossible) to remain stable within any culture. Secondly, for religious beliefs to be advantageous they have to motivate behaviour. And behaviour is only effectively motivated by claims that people think are true. For this reason, false yet useful beliefs must necessarily always remain parasitic upon a basic corpus of beliefs whose utility is a direct result of their accuracy.

We are now almost in the position to see why O'Hear is wrong to worry about the truth of evolutionary theory or of other scientific theories. It is instructive to compare the way religions protect their claims against evidence with the scientific attitude that claims must be subjected to evidence. Undeniably, the scientific attitude is the far less natural one to have (McCauley 2000); the attitudinal conflict between science and religion being far more profound than the one on the level of their respective ontologies. Yet, we have found it to be far better at getting to true beliefs than the religious approach, whose forte may well have been beliefs that gave a selective advantage. Crucially, it is the capacity to apply our cognitive abilities to all parts of the world, including our own mental states, that means we are in the position to even ask whether particular beliefs are advantageous or, indeed, true. Such a critical attitude, however, necessarily requires that we ask whether any particular claim is true, including claims about certain beliefs being advantageous. Ultimately, in the course of cultural evolution, such a critical, scientific attitude has been found to be very effective at spreading itself, even against humanity's natural tendency toward supernatural beliefs. In short, it might be said that O'Hear is right to worry that a predilection for useful falsehoods is natural to humans but that science, with its focus on sometimes uncomfortable truths, has managed to prosper, nevertheless. This success has almost definitely occurred thanks to the enormously increased ability to control our environment science has given us and, just like any evolutionary change that is adaptive in the short-term, may end up leading us to extinction. That possibility is the price of the relativised nature of evolutionary progress, our task being to avoid that outcome.

I do not think that this account of evolved reason will satisfy O'Hear. A clue to why it will not can be found in the part of O'Hear's chapter I quoted at the beginning of this section. He fears that by grasping evolutionary explanations of human reason we lose touch with "ultimate reality." The world of evolution clearly has a "penultimate" feel for him. If that is his point, I can but agree.

There is no reason to think that naturally evolved cognitive abilities should be capable of anything more than, to use Wimsatt's phrase, step-wise and partial approximations to an understanding of the natural world; including, as I have striven to show, an understanding of the human tendency to religious behaviour. Certainly, science does not give us ultimate knowledge—all scientific knowledge is partial and tentative, subject to later revision, rejection or extension. Philosophy's (or any other endeavour's) occasional claims to provide any other kind of knowledge are pure hubris. Were this not the case, the open-ended nature of science would have allowed it to encompass that knowledge. This result will not fully satisfy O'Hear's desire for ultimate reality. I do not think reality needs an adjective, however.

5 Morality

Following his argument against the claim that evolutionary theory can provide a sound basis for an understanding of human reason, O'Hear moves on to a discussion of the supposed ethical implications of evolutionary theory—"let us see what the theory implies about human development" he states. Firstly, he suggests that Darwin's progressivist evolutionary approach provided Darwin (and others, including even Hitler) with an intellectual basis for racism. Secondly, he argues that an approach to social policy founded upon an evolutionary basis would lead to appalling consequences. Neither of these claims stands up under scrutiny. I begin by examining the historical connections between Darwin and Hitler and then consider the question of whether evolutionary theory in any way supports racist or eugenic policies.

Playing the Nazi card is not a line of argument to be used lightly. The attempt to tie something one does not approve of to Hitler's regime is probably the most overused argumentative strategy in non-academic circles, it has even earned its own pig Latin title—*reductio ad Hitlerum*. This does not mean, of course, that all arguments of this form are fallacious. The particular argument that Hitler's virulent racism could be traced back to Darwin has been made many times. Mostly, by creationists (Weikart 2004 is one recent example). O'Hear puts the connection in more measured terms than they do:

> In 2011 it is hard not to be disturbed by Darwin's casual reference to the elimination of endless numbers of lower races, and even more by the way this sort of thinking was taken up by his followers such as Haeckel and von Treitschke, who in turn influenced Hitler.

Given that some of my teaching is done inside the area of the Warsaw ghetto, with the rest being done only a few kilometres from the site of one of the Nazi concentration camps, terms such as "lower races" are definitely something I find disturbing. Yet, nothing much of philosophical interest follows from whether we find something disturbing. If this were all that O'Hear was claiming, therefore, mentioning Hitler would definitely amount to nothing more than an attempt to tar Darwin with the same brush. The claim that lends O'Hear's accusation more weight is the added claim that Darwin influenced Hitler's thinking. For this, however, there is less than no evidence. If we consider Darwin, we must realise that his language, appalling as it sounds in places to modern ears, was merely the language of his times. Indeed, Darwin's attitudes appear to have been admirably liberal compared to his contemporaries. Also, the progressivism that O'Hear claims motivated Hitler is not to be found within Darwin, as I have already shown. In so far as Hitler might have been influenced by progressivist views of biology they are more likely to have their source with Heinrich Georg Bronn who in translating *The Origin of Species* into German inserted his own progressivist views (for an introduction to the literature concerning Bronn's role see Meyer 2009). On the other hand, reading *Mein Kampf* it is clear that Hitler did not believe in Darwinian evolution. Firstly, he never mentions Darwin. Secondly, his very rare use of the term "evolution" is the colloquial one that is synonymous to "progress." Thirdly, his views as to the origin of humanity appear to be traditionally creationist: "For it was by the Will of God that men were made of a certain bodily shape, were given their natures and their faculties" (see Chap. 10 in Vol. 2 of Hitler *Mein Kampf*). One could just as well talk about how disturbing it is in 2011 to read Christ's speech regarding coming not to bring peace but to bring a sword and how this thinking was taken up by his followers including Paul of Tarsus, who in turn influenced Hitler. Given that anti-Semitism in Europe predates evolutionary theory by many centuries and was a major, permanent part of Christian ideology until after the Shoah, such a connection is much easier to defend. Likewise, the traditional Christian picture of a ladder of being is much more conducive to racist interpretations than Darwin's actual theory. Yet, to declare Christianity responsible for Hitler's crimes would also miss the point, even if only partly. On the one hand, the sources of Hitler's individual anti-Semitism are much more complex. On the other hand, the tendency toward xenophobia, unfortunately, appears to be a deeply ingrained human trait whose existence can be understood given its utility in maintaining cohesive social groups in our ancestral environment. Given their shared social role, it is hardly surprising that in many places xenophobia and religion came to be connected in an all-too-often toxic mix.

Rather than getting trapped in a discussion of the historical connections between Hitler and Darwin, however, it is much more useful to consider the basic question of whether the policy implications of evolutionary theory are unacceptable. To do that, however, it is necessary to first ask what Darwin's theory entails concerning how people ought to behave. The answer is simple—it entails nothing. To think otherwise is to fall straight into the naturalist fallacy—an error that would be particularly astonishing for someone of as much an antinaturalist tenor as O'Hear. Evolutionary theory is a scientific description of the causal interrelations between physical events, just like the theory of gravity. And, just like the theory of gravity, it does not tell us how we should react to these facts. It is a fundamental, though common, misunderstanding of evolutionary theory to think that it provides a justification for either eugenics or racialism. One might just as well think that the theory of gravity entails that we should all jump off a bridge, however.

It might be argued, still, that even though evolutionary theory does not actually counsel any particular policy choices, it reveals a choice between policies that are inhuman and options whose consequences are even less palatable. A more charitable interpretation of O'Hear's argument would be, therefore, that evolutionary theory presents us with the unenviable choice of either eugenics or a slow slide into decrepitude due to soft living that allows inferior individuals to survive and pass on their genes. In that case, however, we should simply ask whether evolutionary theory is correct in its implications. If it were, we could not avoid making a choice between them and it would be better to make the choice knowingly, in which case we should be glad that Darwin revealed to us the bad news. If evolutionary theory was incorrect, on the other hand, the problem would not be one of ethics but of scientific accuracy. Either way, the correct response would be to evaluate the evidence and then to let our morality guide us to the best response to the situation revealed by the evidence—not to shoot the Darwinian messenger. As it is, however, only a very naive account of evolutionary theory might be thought to lead to the stark choice outlined above. Of course, due to its fundamental role within the life sciences (including the social sciences), naive accounts of evolution abound in popular culture and are often used for various aims that have nothing to do with actual science. Yet we must never confuse the two. The real science of genetics shows that eugenics programmes will almost definitely fail since most genetic disorders are recessive and recessive genes are common. So common that a eugenics programme which would seek to eliminate "bad" genes would have to eliminate all of us since we all carry some such recessive genes (and this does not even take into further complications such as genes that are beneficial in their recessive form). Also, real genetics shows that notions of race are culturally constructed rather than reflecting any underlying biological reality. There is no non-arbitrary way to distinguish particular large groups of people who are more closely genetically related.

6 A World for People

Underlying O'Hear's objections seems to be a fear that, were Darwin right, the world would be shown to be immoral. Far from it, however, it is evolutionary theory that has explored selective processes that explain how morality has come to exist. Robert Trivers (1971) has developed the notion of reciprocal altruism which helps us to understand why organisms enter into stable mutualist relationships. W.D. Hamilton's (1968) work on kin selection has provided a formalised account explaining the willingness of some individuals to even give up their lives for the good of their family. Elliot Sober and D.S. Wilson (1998) have used a multilevel selection model to argue that co-operation between group members may be explained in terms of the selective advantage it grants on the group level. At the same time, it is zoology that shows the world of non-human animals to be a fecund source of examples of individuals acting upon instincts that it would be mere human chauvinism not to call noble. The objections that all of this work has nothing to do with the way humans choose to sacrifice their own good for that of others misses the point. People are capable of real altruism, unmotivated by possible future gain or whether the person helped is a member of the community or the family. But that is a matter of the psychological mechanism—in evolutionary terms: a proximate explanation. The explanations given by Trivers and the others, however, work one level up, being concerned with the question of whether particular kinds of behaviour would spread due to selective advantage. They do not necessarily concern themselves with the mechanisms which might cause that behaviour. An ability of empathise with others and a willingness to help them without much thought for oneself, such as people sometimes exhibit, would be a fine mechanism to achieve such behaviour in the human ancestral environment. It isn't just humans that do not consider the evolutionary consequences of their actions, after all. Having evolved a moral sense and the consciousness to consider it, however, we are no longer the blind pawns of evolution, even though we are still subject to its laws. A personal example seems most appropriate at this juncture. I have at this time two daughters, whom I love and treasure more than anything else in this world. The evolutionary reasons for why a mammalian parent should so cherish their offspring are well known to me and I am certain they apply, just the same, in my case. Yet, this fact no more changes my love than the fact that the Mona Lisa is made of paint strokes on a canvas alters its beauty. If anything, it makes it all the more precious, for the reasons suggested in O'Hear's quote from Monod and so powerfully spelled out in Matthew Arnold's "Dover Beach." There is no universal progress in evolution and there are no universal values. Yet, there is a real morality that is tied to our way of life as fragile, evolved, and social beings. The facts delimit our options but do not free us of the need to make choices.

O'Hear's own alternative is to call forth upon the strong anthropic principle, i.e. it would have been so improbable for us to come about by accident that the reason the universe exists must be to create us. But, it is simultaneously too strong to be plausible and too weak to achieve what he requires (as numerous discussions of its use by creationists have suitably shown). On the one hand, to make the principle plausible, O'Hear would have to be able to sensibly talk about all of the possible universes and to claim that he is able to know in which of them intelligent life would evolve. Yet, it is not even clear what sense of "possible" is being referred to here, since it cannot be physical possibility that is under consideration. On the other hand, if we do accept the principle, there does not seem to be any principled way to distinguish between different senses of "us." Of course, normally the principle is thought to refer to intelligent beings but why stop there? It is more improbable that the intelligent beings be humans and it is even more improbable that the intelligent beings be the particular set of individuals currently alive. Does this mean that the reason the universe was created was so that I should exist? The hubris is breathtaking. At the same time, even accepting the principle in its more usual form, all we have is the conclusion that intelligent life is the reason for the existence of this universe. Any step taken beyond this conclusion, beyond "purely physicalistic terms," is groundless—a skyhook—and therefore incapable of providing a real explanation, as has already been noted previously. Even with the strong anthropic principle in place, one needs to explain how intelligent life came to develop, and either intelligent life can be explained in terms of processes of emergence or it can not be explained at all. Even if the strong anthropic principle was, *per impossibile*, thought to explain the *why* of intelligent life, you still need evolution to explain the *how*. Thankfully, the theory of evolution is perfectly capable of doing that, as well as helping to explain why anthropocentric ideas that treat reasons as more fundamental than causes, such as the strong anthropic principle, will be intuitively attractive to humans.

7 Conclusion

The human condition—as revealed by scientific investigations and rendered clear by evolutionary theory—is not that of a fortunate race at the peak of creation, basking in a womb-like world made with it in mind. Darwin did not attempt to present humanity in this way, eschewing the progressivism that plagued the nineteenth century and which still afflicts the imaginations of many. The theory he did develop has proved extraordinarily powerful, being able to explain a range of phenomena from across what had previously been thought to be widely disparate sciences. Indeed, it is only now with the breaking down of the wall between the *Geisteswissenschaften* and the *Naturwissenschaften*, made possible by recent work in evolutionary explanations of human behaviour, that the true potential of Darwin's thought is finally being realised. Vitally, this work makes it possible to properly understand emergent phenomena such as cognition and value that previously could not be accounted for as anything but primitive elements of reality. Most fascinatingly, perhaps, it makes it possible to understand the very tendencies that make it difficult for human minds to fully accept our evolutionary heritage. As such, Darwin's theory of evolution stands at the centre of naturalised philosophy.

Acknowledgments

Some of the ideas presented here were developed during the discussion at the "Knowledge, Value and Evolution" meeting in Prague: my thanks to the participants. A draft of this essay benefited from Jonathan Knowles' comments.

References

Barrett, J. (2000). Exploring the Natural Foundations of Religion. *Trends in Cognitive Sciences* 4: 29-34.
Bickhard, M. H. and Terveen, L. (1995). *Foundational Issues in Artificial Intelligence and Cognitive Science: Impasse and Solution.* Amsterdam: Elsevier.
Bishop, M. A. and Trout, J. D. (2004). *Epistemology and the Psychology of Human Judgement.* Oxford: Oxford University Press.
Darwin, C. (1859). *The Origin of Species.*
Dawkins, R. (1986). *The Blind Watchmaker.* New York: Longman.
Dennett, D. (1995). *Darwin's Dangerous Idea: Evolution and the Meanings of Life.* New York: *Simon & Schuster.*
de Sousa, R. (2007). *Why Think? Evolution and the Rational Mind. Oxford: Oxford University Press.*
Diamond, J. (1997). *Guns, Germs, and Steel: The Fates of Human Societies.* New York: Norton.
Feltz, A. and Bishop, M. (2010). The Role of Intuition in Naturalized Epistemology. In *Beyond Description: Normativity in Naturalised Philosophy.* M. Milkowski and K. Talmont-Kaminds (eds.). London: College Publications.
Haack, S. (2005). Not Cynicism but Synechism: Lessons from Classical Pragmatism. *Transactions of the Charles S. Peirce Society* 41 (2): 239-53.
Hamilton, W. D. 1963). The Evolution of Altruistic Behavior. *American Naturalist* 97: 354-356.
Hooker, C. (1991). Between Formalism and Anarchism. In *Beyond Reason.* G. Munévar (ed.), Amsterdam: Kluwer.
Hume, D. (1779). *Dialogues Concerning Natural Religion.*

Laland, K. L. and Brown, G. R. (2002). *Sense and Nonsense: Evolutionary perspectives on Human Behaviour*. Oxford: Oxford University Press.

Lewontin, R.C. (1982). *Human Diversity*. New York: Scientific American/Freeman.

MacArthur, R. H. and Pianka, E. R. (1966). On the Optimal Use of a Patchy Environment. *American Naturalist* 100: 603-609.

McCauley, R. N. (2000). The Naturalness of Religion and the Unnaturalness of Science. In *Explanation and Cognition*. F. Keil and R. Wilson (eds.). Cambridge, Mass.: The MIT Press: 61-85.

McKay, R. T. and Dennett, D. C. (2009). The Evolution of Misbelief. *Behavioral & Brain Sciences* 32.6: 493-561.

Meyer, A. (2009). Charles Darwin's Reception in Germany and What Followed. *PLoS Biol* 7(7): e1000162.

Parker, A. (2003). *In the Blink of an Eye*. Cambridge, Mass.: Perseus Books.

Pearce, J. M. (1997). *Animal Learning and Cognition*. Hove, UK: Psychology Press.

Simon, H. (1996). *The Sciences of the Artificial*. 3rd Ed. Cambridge, Mass.: The MIT Press.

Sober, E. and Wilson, D. S. (1998). *Unto Others: The Evolution and Psychology of Unselfish Behavior*. Cambridge, Mass.: Harvard University Press.

Talmont-Kaminski, K. (2009a). The Fixation of Superstitious Beliefs. *Teorema* 28 (3): 81-96.

_____ (2009b). Effective Untestability and Bounded Rationality Help in Seeing Religion as Adaptive Misbeliefs. *Behavioral & Brain Sciences* 32 (6): 536-537.

_____ (forthcoming). *In a Mirror, Darkly: How Superstition and Religion Reflect Rationality*.

Trivers, R. L. (1971). The Evolution of Reciprocal Altruism. *Quarterly Review of Biology* 46: 35-57.

Weikart, R. (2004). *From Darwin to Hitler*. New York: Palgrave-MacMillan.

Wilson, D. S. (2002). *Darwin's Cathedral: Evolution, Religion, and the Nature of Society*. Chicago: The University of Chicago Press.

Wimsatt, W. (2007). *Re-engineering Philosophy for Limited Beings: Piecewise Approximations to Reality*. Cambridge, Mass.: Harvard University Press.

4

Reply to Konrad Talmont-Kaminski

Anthony O'Hear

I am grateful to Konrad Talmont-Kaminski (T-K) for his reply to my chapter, and grateful to the editors of this book for the chance to reply to the reply. I shall do so briefly, focusing on what I take to be the main points of disagreement.

1 Darwin Himself

T-K accuses me of philosophical irrelevance in discussing a historical figure (C. Darwin), rather than current evolutionary science. It all depends on what one is trying to do in a philosophical discussion. Many of the historical Darwin's ideas have permeated contemporary thought, and continue to be part of contemporary high culture; for this reason alone, therefore, their basis and origin needs unmasking. I would point here particularly to the naturalistic (very Darwinian) thought that human behaviour can be revealingly analysed in Darwinian or at least in evolutionary terms, and also to the premise, shared by a good proportion of atheists and by many religious believers alike, that Darwin's theory (and I do mean Darwin's) has somehow shown religious belief to be untenable. I have written a whole book on the first topic (see O'Hear 1997), and much of the chapter we are discussing here is about it, too. As T-K does raise the point, though, I will clarify my attitude to creationists and anti-creationists, so to speak. Creationism is a largely negative strategy (pointing to difficulties in evolutionary theory), and vulnerable to the extent that were biologists to sort out the difficulties, creationism would thereby become redundant. More important, though, even if we were to grant its analysis of evolutionary theory, creationism would tell us very little about religion properly speaking. For it makes God into a quasi-scientific being among beings, invoked at points where empirical science seems to be in difficulty—what used to be called a god of the gaps—rather than the ground of all being, wholly beyond the type of consideration adduced by those who argue (on both sides) about what has come to be called creationism. Correctly considered, the God of religion is, by contrast, impervious to anything Darwin and his followers might have to say about the explanatory adequacy of the mechanics of evolution.

2 Progress

To return to T-K's worries about my focus on Darwin, where he and I would part company is that I think that metaphor and rhetoric are not ineliminable even from the driest science, and particularly not where that science is held to adumbrate or support a world view, which Darwinism and evolutionary theory most assuredly do. So while I am as aware as T-K that, in certain moods Darwin himself, and some of his more recent successors such as Lewontin and Gould, would deny any progressivist sense to the theory of evolution, in practice it is very hard, if not impossible, for anyone, including Darwin himself, not to conceive of it in terms of an *ascent* from the simpler to the more complex, and from the less to the more intelligent, more sensitive, more morally complex, etc. This is why I call "progressivism" and what I say about the Creator or an ersatz-Creator, a *tension* in Darwinian and, I would hazard, in evolutionary thinking more generally. If the tension doesn't exist any more, and if all true scientists are all now strict "descent with modificationists," as Gould would have wished, not believing in ascent or anything like that, then well and good; what I have to say is indeed about no more than a historical curiosity. But that is not how evolution is pictured in the popular mind and in popular accounts, or, as I show, how it was pictured (sometimes) by C. Darwin himself. So there is still something which needs to be said here, maybe (dare I say?) along the lines of my chapter. And even contemporary scientists need to be aware of the problems involved in a progressivist account of evolution, as the temptation at least occasionally to think, and even more to talk in that way is well-nigh irresistible.

3 The Naturalistic Fallacy

T-K accuses me of committing the naturalistic fallacy, in that I appear to be drawing ethical and political consequences from Darwin's theory. Actually it is Darwin who draws the consequences, not me, and it is hard to see how people who write books with titles like *The Selfish Gene* or *Darwin's Dangerous Idea*, despite protestations to the contrary, are not doing something rather similar. However, I actually think that T-K's invocation of the naturalistic "fallacy" as genuinely a fallacy, and hence as something at all costs to be avoided is too quick, certainly in Darwin's case. In *The Descent of Man* Darwin labours very hard to describe human nature, and in what is to-day called evolutionary psychology something very similar takes place. If there is any validity in what is said in these places, it is hard not to see them having a bearing on ethics and politics (to put it no stronger). After all, our ethical and political practices should surely take the nature of human beings into account. My quarrel with Darwin and the others is not that they ground morality and politics in conceptions of human nature; it is rather whether the conceptions of human nature they work with are adequate.

4 A Man of His Time

Darwin certainly advocated what has come to be called social Darwinism and what I dubbed "race and species progressivism." T-K, while conceding this, says that we have to realise that Darwin was of his time, when lots of people thought this sort of thing. This may or may not be true, but it is beside the point. The worry I have is not about what Darwin himself thought; it is about whether these views (including their eugenicist consequences) in some way derive from the theory. Despite the efforts of modern neo-Darwinists to show that at least indirectly they do not, it is hard to see otherwise. For the theory (survival of the fittest) says that the fittest will survive and the weakest be eradicated, and that if we want the species to continue to progress (!), we must not interfere (too much) with this process. Looked at like this, it seems that it is not so much that Darwin is a man of his time, as that the theory is a theory of its and Darwin's time. And whether or not Hitler believed in the Darwinian evolution of species, he certainly did believe that the fittest should be encouraged and the unfit discouraged, as did very many on the political right AND left in the first part of the twentieth century (eugenics), in many cases, including some who did influence Hitler, explicitly drawing on the theory of evolution for support. Now, of course, I know that "is" does not strictly imply "ought;" so even if we are Darwinians we are not obliged to contemplate policies which fail to cherish and protect the weak. An initial problem for a compassionate Darwinian, who wants to resist the theories unpalatable implications—if he thinks that his account of human nature is anything like a complete one—is to explain where true compassion and true altruism are going to come from. T-K would no doubt talk about game theory and reciprocal altruism at this point, to which all I can say here is that reciprocal altruism is close to true altruism as paid-for love is to true love (not a bad analogy, as it happens). But could it be that through some process of evolutionary development to encourage in-group co-operation, reciprocal "altruism" has somehow come to engender in us feelings which break free of their game-theoretic basis and take us to true altruism? In the doubtless lamentable absence of evidence of how pre-human and proto-human societies actually developed, nobody can know the answer to this question; but what is clear is that if these truly altruistic feelings started getting a hold in a population, the Darwinian process of improvement through struggle would be severely impeded, to put it no stronger, which is, of course, what Darwin and the eugenicists were worried about, because such a society would in due course fare badly against a stronger, more resolute competitor—which is the second problem a Darwinian who wishes to resist eugenicist conclusions will have to deal with. An alternative view (which I would urge) would be that the Darwinian picture of human nature is at best partial; and that, the theory of natural selection notwithstanding, human beings are not constituted only by dispositions which have clear adaptive advantage in the struggle for survival and reproduction.

5 Cognition

I actually agree with quite a lot of what T-K says about our cognitive interests and faculties. Our disagreement would be on the jump from what he calls bounded rationality, where our cognitive faculties work adequately enough in a particular environment, to what I would conceive our actual situation to be, when we enquire into all sorts of things way beyond the bounded environment of our putative ancestors and using a form of rationality which is committed to discovering the true rather than eliciting the useful. In a similar vein to my treatment of a supposed jump from reciprocal altruism to true altruism, I would suggest that in our cognitive development we have come to adopt an attitude which values truth for its own sake, irrespective of advantage, and also to value the pursuit of enquiries which seem to be worth pursuing for their own sakes, again irrespective of survival or reproductive advantage. This suggests at the very least that we have broken free from the constraints of evolutionary accounts and explanations, and entered a new plateau of existence, whatever the ladder by which we reached it (whether crane or sky-hook). So whatever evolution might tell us about origins here, it may not tell us much about how we go now. In discussing the transition from a socially cohesive society sustained by religious myth to the adoption of scientific rationality (which I presume undermines the myth), T-K says "that a critical scientific attitude has been found very effective at spreading itself." But to put it like that suggests that the critical scientific spirit is just another myth, on the same epistemic level as the rest, but just socially and institutionally more powerful than the superstitions it displaced, as maybe Darwinism is to-day replacing six or however many day creationism (if indeed it is, outside the circles in which T-K and I move). Where, in this quasi-evolutionary account of an idea successfully spreading itself, do reason and truth get a foothold? Once beliefs are examined for their truth and rationality, rather than for their evolutionary advantage, we are on a new level, where evolutionary explanations and accounts seem to be focusing on the wrong things, especially as there can be no general assumption that the truth is going to be advantageous in either the short or the long term or that falsity cannot be at times highly advantageous. But we pursue truth and an astringent rationality nonetheless (or think we should), even in areas where there could be no conceivable pay-off.

6 The Anthropic Principle

I do not know whether we should accept the anthropic principle. In a way in my chapter I was simply entertaining it as an alternative picture to the Monod-Russell view (metaphor, again, if you like). I do feel that both life and consciousness are very hard to account for on physicalistic grounds, and a picture which has them inherent in the beginning of things might make more sense than one which has them inexplicably and randomly arising on just one planet. I never, though, meant to imply that on the anthropic principle human existence would be, as it were, the omega point of the universe, and I thank T-K for pointing out that I may have given that impression. Far more likely on an anthropic view would be lots of life in many parts of the universe, and many degrees of intelligence etc., with no presumption that human life is at the highest level. However, I am worried by T-K's observation that the anthropic principle commits me to the very progressivism I objected to in Darwin. I think, though, that I might be able to say that the progressivism which I object to in Darwinism is one which is based on the survival of the fittest, whereas I suggested that some version of the anthropic principle might be conducive to a view which stressed a community and a mutual belonging throughout the universe. But, even with this exit strategy, I do not pretend that it is easy to see progress of any sort in human affairs, to look no further, and I am indebted to T-K for pointing out that the picture I am toying with may well have an implication of this sort.

References

O'Hear, A. (1997). *Beyond Evolution: Human Nature and the Limits of Evolutionary Explanation*. Oxford: Oxford University Press.

5
Fodor vs. Darwin: A Methodological Follow-Up
Lilia Gurova

In a series of publications, which appeared in the last few years,[1] Jerry Fodor has launched an attack on what many believe is the core of Darwinian theory of evolution–the theory of natural selection. Fodor complains that the theory of natural selection "can't explain the distribution of phenotypic traits in biological populations" (Fodor 2008a, 11) and his main argument, slightly simplified[2], is the following: In order to play its explanatory role properly, the theory of natural selection must rely on "nomologically necessary generalizations about the mechanisms of adaptation as such" (Fodor 2008a, 23). There are not good candidates for such "nomologically necessary generalizations," therefore, the theory of natural selection cannot explain what it is supposed to explain.

Not surprisingly, Fodor's attack provoked a strong, mostly negative, reaction.[3] Fodor's critics have complained that he does not really understand how evolutionary biology works.[4] They have insisted both that his main argument

---

[1] See (Fodor, 2007a), (Fodor, 2007b), (Fodor, 2008a), (Fodor & Piatelli-Palmarini, 2010).

[2] The original form of Fodor's "putative argument" is the following:

(i) Explaining the distribution of a phenotypic trait in a population would require a notion of "selection for" a trait. "Selects for ..." (unlike "selects... ") is opaque to substitution of co-referring expressions at the "..." position.

(ii) If T1 and T2 are coextensive traits, the distinction between selection for T1 and selection for T2 depends on counterfactuals about which of them *would be* selected in a *possible* world where the *actual* coextension doesn't hold.

(iii) The truth makers for such counterfactuals must be either (a) the intensions of the agent that affects the selection, or (b) laws about the relative fitness of having the traits.

(iv) But:

Not (a) because there is no agent of natural selection.

Not (b) because considerations of contextual sensitivity make it unlikely that there are laws of relative fitness ("laws of selection").

(v) QED. (Fodor 2008a, 11)

[3] See Sober (2008), Godfrey-Smith (2008), Dennett (2008), Block & Kitcher, (2010), and Ruse (2010).

[4] Block and Kitcher (2010), for example, say that Fodor's argument is "biologically irrelevant," Dennett (2008) blames Fodor for relying too much on a "caricature of scientific practice," and

is unsound and that his central claim is false. I can generally agree with the first part of their criticism: Fodor's "putative argument" does rely on controversial premises that make it unsound.[5] However, I don't think that Fodor's critics have succeeded in their attempts to refute his central claim. The refutation strategy that most of them have undertaken is to show examples of successful evolutionary explanations by natural selection. In what follows I analyze two of these examples, which have been suggested by the philosophers of biology Elliott Sober and Peter Godfrey-Smith. The analysis reveals that:

(1) In both examples the evolutionary explanations by natural selection rely on additional empirical hypotheses; these hypotheses might be true but they also might be false. This observation is in tune with what Fodor has said about the successful evolutionary explanations: they are such because evolutionary biologists have at their disposal more than the theory of natural selection. Thus the theory of natural selection should be only partially credited with the explanatory success of such explanations.

(2) In both cases alternative non-evolutionary explanations can be found that fit the same empirical data and no reason has been given why these alternative explanations should be ignored a priori as inferior.

(3) The observations (1) and (2) stand against the claim that theory of natural selection is the only legitimate *explanans* for the distribution of phenotypic traits. This does not mean, of course, that natural selection does not play any explanatory role or that the theory of natural selection is a false theory (as Fodor is inclined to argue). This only means that there is indeed a problem of understanding the proper explanatory role of natural selection and that this problem is not only Fodor's problem. In the conclusions of this chapter an outline will be given of what should be admitted in order to get to a better understanding of the explanatory role of the theory of natural selection.

1 The First Example: Fisher's Sex Ratio Model

According to Sober (2008), what Fisher mathematically inferred on the basis of his model is a good candidate for a law, which explains/predicts the 1:1

Ruse (2010) states explicitly that what one can only say about Fodor's claims concerning the theory of natural selection is "that this is a misunderstanding of the nature of science."

[5] I, for example, agree that "nomologically necessary generalizations" are not necessary conditions for producing good explanations.

sex ratio which is observed in most species: "If producing equal numbers of sons and daughters and producing more daughters than sons are the alternative reproductive strategies that a parent might follow in a randomly mating population, and if the cost of rearing a son is the same as the cost of rearing a daughter, then there will be selection for following the first strategy and against following the second" (Sober 2008, 45).

Fisher's argument (see Fisher 1930) is going in the following direction. If we assume for a while that the males in a given population are less in number than the females, the average contribution of each male to the total reproductive value (the offspring) of this population will be, for obvious reasons, higher than the average female contribution to the same reproductive value. That means that the parents who possess the natural tendency to produce more sons than daughters will create a higher contribution to the total reproductive value of the population. Thus their genes will spread more than the genes of those who are not genetically disposed to have more sons than daughters and this tendency will last until the moment when the contribution to the reproductive value of males and females become equal and this will happen when they become equal in number.

Fisher's principle has often been celebrated as one of the most remarkable achievements of evolutionary biology (Edwards 1998). This is so not only because it successfully explains the observed 1:1 sex ratio in most species but also because it implies the empirically confirmed prediction that if in a given population rearing sons is more "expensive" than rearing daughters, there will be "selection for" producing smaller number of sons than daughters.[6]

Despite the broadly admitted explanatory success of Fisher's principle, two things about its use must be stressed.

First, Fisher's principle only works as a supporting selectionist explanation of sex ratio if we assume that there is a genetically inherited disposition to produce more male or more female births. This is an empirical conjecture which has not been yet confirmed for most species. (For sure, at the time when Fisher published his book there had not been any evidence for the existence of such inheritable dispositions). That means that the evolutionary explanations of sex ratio based on Fisher's principle are at best tentative explanations.

Second, those who seem to neglect the tentative character of Fisherian sex ratio explanations have probably never asked seriously this question: is it possible to explain what Fisher's principle explains without assuming the influence of any selection pressure? Because if they had asked this question they would have easily discovered that the answer is "yes" for both the 1:1 ratio pre-

---

[6] This prediction has been well confirmed by some recent studies of sexually dimorphic Hymenoptera (see Seger & Stubblefield 2002).

diction when daughter and sons cost equally and the prediction that less sons will be given birth if rearing a son is twice as expensive as rearing a daughter. The ratio 1:1 can be easily explained by just assuming that sex allocation is a random process. Then in the case of two sexes, the prediction is exactly about equal number of male and female births. Let's suppose that for certain reasons (a dreadful war, or a strange male-killing pandemic disease) the number of males is crucially reduced. The ratio 1:1 will be restored immediately in the next generation just because of the randomness of the process of sex allocation. What about the asymmetry between male and female births when rearing a son is most costly? It also allows a simple explanation by just assuming equally probable male and female births and assuming also that all female parents can make (and do make) a limited investment in rearing children.

Let me clarify this by the following example. Let's assume that rearing a boy is twice more costly than rearing a girl and that the maximal investment which each mother can make is for four daughters (or two sons). Then in a situation of a random sex allocation we have the following 8 possible cases: (The strings below represent the possible sequences of births, "S" stands for giving birth to a son, and "D" stands for giving birth to a daughter[7]:

(1) S S

(2) S D S

(3) S D D

(4) D S S

(5) D S D

(6) D D S

(7) D D D S

(8) D D D D

If all 8 cases are equally probable, in a population obeying the stated above conditions, there will be 10x male births vs. 15x female births. Thus there will be a strong bias (2:3) toward less male than female births and this will happen independently of any selection pressure. Notice, that no assumption about inheritable dispositions to have more sons than daughters or vice versa is needed in this explanation.

---

[7] It is seen that in cases (2), (4), and (7) the investment exceeds the limit. This happens because before the last birth the mother still has resources for one more daughter but instead of a daughter she gives a birth to a son. Excluding the last births of these cases, however, will not change the general result.

It is well known that what is broadly called "Fisher's Principle" is not Fisher's invention. The roots of the underlying argument can be traced back to Darwin's first edition of *Descent of Man* (1871), where he presented a similar, although more obscure, line of reasoning. For many, it is still a curious fact, however, that Darwin dropped his sex ratio evolutionary explanation from the second (quite more broadly known) edition of the book, providing the following explanation, which one can find also cited by Fisher (1930):

> In no case so far as we can see, would an inherited tendency to produce both sexes in equal numbers or to produce one sex in excess, be a direct advantage or disadvantage to certain individuals more than to others; [...] I formerly thought that when a tendency to produce the two sexes in equal numbers was advantageous to the species, it would follow from natural selection, but I now see that the whole problem is so intricate that it is safer to leave its solution for the future. (Darwin 1874, 399)

There are different explanations of Darwin's decision to abandon what has been later recognized by the mainstream evolutionary biologist as "the right explanation." But in light of alternative non-selectionist explanations of the chief sex ratio phenomena, presented above, Darwin's cautiousness does not look that strange or naive.

I am far from calling for a radical revision of current models of sex ratio dynamics. I do admit that these models are a great success of modern biology insofar the existence of many important correlations which have been predicted by these models (for example, correlations between parental investment, sex rates, and mating schemas) have been also empirically confirmed.[8] But it is a well known fact that correlation does not imply causal connection. In the case of the sex ratio models, the correlations do not imply any causal "selection for" particular observed sex ratios. On the contrary, what I hopefully have been able to demonstrate, many of those empirically confirmed correlations allow non-selectionist explanations.

Let me summarize what the Fisher's Principle example reveals about the explanatory role of theory of natural selection. Two important observations are to be stressed. First, the principle plays its explanatory role only in conjunction with the empirical conjecture that there might be inheritable dispositions for having more sons than daughters or vice versa. Second, the phenomena, which this principle explains allow alternative non-selectionist explanations. This means that further research is needed in order to decide whether the evolutionary explanations describe the actual course of events better than their non-evolutionary rivals. Before having the results of this research, one cannot

---

[8] For a recent review of research in this field see Hardy (2002).

conclude that natural selection is a necessary part of any proper explanation of the distribution of all observable phenotypes.

2 The Second Example: the Evolutionary Explanation of Aging

Godfrey-Smith's (2008) discussion on the evolutionary explanations of aging has been provoked by Fodor's complain that these explanations are essentially *post hoc*:

> it's often suggested that the reason there are so many diseases of old age is that creatures can't compete for representation in the gene pool once they become infertile. But then, why didn't selection just increase the length of the fertile period? (Fodor 2008a, 13)

In reply, Godfrey-Smith presents two of the most influential models of the evolution of aging, which, he notes, are not incompatible. Both models aim to describe how the evolution brings to the phenomena of aging in a population, which at the beginning did not show any senescence.

According to the mutation accumulation theory (Medawar 1952), aging is a by-product of natural selection which has successfully "selected against" detrimental mutations that manifest their effects in early age (the individual possessing such mutation either die before achieving reproductive age, or do not reproduce because of different malfunctions) but has failed to "select against" any harmful mutations which effects are switched on at later age. Mutation accumulation theory has produced several testable hypotheses which have been confirmed. For example, it predicted successfully that inbreeding depression should increase with age (Hughes *et al.* 2002). However, this theory does not produce correct predictions for populations that are free of predators (Bowles 2000). So, the assumption that there have been enough natural accidents to reduce the number of the older individuals in the initial no senescence population is vital for the explanatory success of mutation accumulation theory.

The antagonistic pleiotropy theory of aging (Williams 1957) seems to remedy the defects of Medawar's model but only on the cost of a new assumption that some genes may affect more than one trait in an organism (pleiotropy) and that these connected traits may play antagonistic roles with respect to fitness. According to this theory, aging appears because evolution has "selected for" traits which are advantageous to reproductive success earlier in life but which are genetically connected to traits which become harmful at later age. The theory predicts that genes that increase the early age productivity will in the same time lead to speeding-up the process of aging. The evidence for this hypothesis, however, is controversial (see Economos & Lints 1986).

As in the case of Fisher's sex ratio principle, the evolutionary explanations of aging also have their non-evolutionary rivals—to mention only the programmed aging theory, neuroendocrine theory, wear-and-tear theory, immune system theory etc.—see Pankow & Solotoroff (2007) for a review. Some of these theories demonstrate no less explanatory success than the evolutionary theories, according to Bell (1984) and Le Bourgh (1998).[9]

So, given that the explanations of aging by natural selection rely on additional hypotheses which have not been yet well confirmed, and given that rival non-evolutionary explanations have the same explanatory success, are there any reasons to claim that the theory of natural selection is the only legitimate *explanans* for aging? The answer of this question, I think, is obvious.

3 Conclusions

The analysis of two examples of evolutionary explanations by natural selection reveals that these explanations are in the best tentative hypotheses, which are not directly inferred from the theory of natural selection but rely in an essential way on additional empirical conjectures which are to be tested independently. In this sense Fodor's claim that theory of natural selection cannot, on its own, explain the distribution of phenotypic traits is correct. It seems to be correct also because there are alternative non-evolutionary explanations for the distributions at least of some phenotypic traits and these alternative explanations cannot be simply ignored as inferior. The practice of ignoring the non-evolutionary alternatives without paying attention to how plausible are they and what is their explanatory power is typical for Darwinian fundamentalism. But Darwinian fundamentalism, which might be indeed harmful for science, must be distinguished from Darwinism. Darwin himself was quite cautious to warn that natural selection is just one of the many forces in the process of evolution.

Darwinian fundamentalism builds on a deep misunderstanding of the proper explanatory role of the theory of natural selection. That's why getting to a better understanding of how the theory of natural selection contributes to the evolutionary explanations is vital for the successful overcoming of harmful selectionist fundamentalism. Perhaps a lot of work is to be done in this direction but it may suffice as a beginning to take seriously the following. Natural selection is a negative force. That means that literally it only "selects against." "Selection for" is a metaphor for what has survived the "selection against." But

---

[9] But it should be noticed in the same time that the evolutionary biologists also complain that their theories have been almost completely ignored by the representatives of the mainstream gerontology (see Rose *et al.* 2008).

the explanation that a particular phenotypic trait is there because it has not been selected against is at best a partial explanation. It misses the essential complementary story about how this trait was brought to life and what made it to flourish. And this story must necessarily rely on additional hypotheses that in themselves have nothing to do with selection. However, the success of the evolutionary explanations "by natural selection" depends crucially on the truth of these additional hypotheses. I am completely aware that what I just have said is not news but the reaction against Fodor's attack on what he has (wrongly) recognized as "Darwinism" has convinced me that it deserves to be stated again.

Acknowledgments

I thank the organizers of the conference "Knowledge, Value, Evolution" (23-25 November, Prague) Juraj Hvorecký and Tomáš Hříbek for giving me the chance to deliver an early version of this chapter and to receive a useful feedback on its main ideas. I am especially indebted to Malcolm Forster who read the last version of the manuscript and suggested many helpful corrections.

References

Bell, G. (1984). Evolutionary and Nonevolutionary Theories of Senescence. *American Naturalist* 124: 600-603.
Block, N. and Kitcher, P. (2010). Misunderstanding Darwin. *Boston Review*, March/April.
Bowles, J. (2000). Shattered: Medawar's Test Tubes and Their Enduring Legacy of Chaos. *Medical Hypotheses* 54: 326-339.
Darwin, C. (1871). *The Descent of Man and Selection in Relation to Sex*. London: John Murray.
Darwin, C. (1874). *The Descent of Man and Selection in Relation to Sex*. 2nd Ed. London: John Murray.
Dennett, D. (2008). Fun and Games in Fantasyland. *Mind & Language* 23, 1: 25-31.
Economos, A. and Lints, F. (1986). Developmental Temperature and Life-span in Drosophila Melanogaster: Pt. 1. *Gerontology* 32: 18-27.
Edwards, A. (1998). Natural Selection and the Sex Ratio: Fisher's Sources. *American Naturalist* 151: 564-569.
Fisher, R. (1930). *The Genetical Theory of Natural Selection*. Oxford: Clarendon Press.
Fodor, J. (2007a). Against Darwinism. In *Proceedings of EuroCogSci: 07*. E. Vosniadou, D. Kayser and A. Protopopos (eds.). New York: Lawrence Erlbaum.
_____ (2007b). Why Pigs Don't Have Wings. *London Rev. of Books*, 29, Oct. 18th., 20.
_____ (2008a). Against Darwinism. *Mind & Language* 23 (1): 1-24.

_____ (2008b). Replies. *Mind & Language* 23 (1): 50-57.
_____ and M. Piatelli-Palmarini (2010). *What Darwin Got Wrong*. London: Profile Books.
Godfrey-Smith, P. (2008). Explanation in Evolutionary Biology: Comments on Fodor. *Mind & Language* 23 (1): 32-41.
Hardy, I. (ed.) (2002). *Sex Ratios. Concepts and Research Methods*. Cambridge: Cambridge University Press.
Hughes, K., Alipaz, J., Drnevick, J. and Reynolds, R. (2002). A Test of Evolutionary Theories of Aging. *Proc. Nat. Acad. Sci.* 99: 14286-14291.
Le Bourgh, E. (1998). Evolutionary Theories of Aging: Handle with Care. *Gerontology* 44: 345-348.
Medawar, P. (1952). *An Unsolved Problem of Biology*. London: H. K. Lewis.
Pankow, L. and Solotoroff, J. (2007). Biological aspects and Theories of Aging. In J. Blackburn and C. Dulmus (eds.). *Handbook of Gerontology*. N.J.: John Wiley and Sons.
Rose, M., M. Burke, P. Shahresteni and L. Mueller (2008). Evolution of Aging Since Darwin. *Journal of Genetics* 87: 363-371.
Ruse, M. (2010). Origin of Specious. *The Boston Globe*. February 14.
Seger, J. and Stubblefield, W. (2002). Models of Sex Ratio Evolution. In Hardy 2002, 2-25.
Sober, E. (2008). Fodor's *Bubbe Meise* Against Darwinism. *Mind & Language* 23 (1): 42-49.
Williams, G. (1957). Pleiotropy, Natural Selection and the Evolution of Senescence. *Evolution* 11: 398-411.

6

Darwin's Inference of Origins

Aviezer Tucker

As the title of his *magnum opus* attests, one of the purposes of Darwin's theory of evolution was the inference of the origins of species. This chapter attempts to understand how. How did Darwin infer the origins of various groups of species? What were his methodology and theoretical assumptions?

I argue that Darwin's inference of origins was modular, in three consecutive stages. First, Darwin proved that some homologous, information preserving, similarities between species, are more likely given common causes than given separate causes, without specifying the properties of the common causes. By contrast, the convergence of evolutionary beneficial traits, homoplasy, tends to be more likely given separate causes. Second, if similarities between species are proven more likely given some common causes in the first stage, five types of common cause causal-information transmitting genealogical models are possible. Darwin had to try to find out which of the five possible nets makes the similarities most likely: a single distinct ancestor species; several ancestral species; a single ancestor species that is also a member of the set of species that was determined to have had a common cause; several ancestor species that are also members of the set whose similarities are explained who then interbred; or, finally, the genealogical map can combine type 1 or 2 models with type 3 or 4, i.e., the descendents of a common ancestor species or species may then interbreed with each later. Third, once a particular causal model is determined, it is possible to attempt to infer the properties of the common cause(s), the origins. As Sober (2002) noted, the inference of tree typology needs to be distinguished from the inference of character states of the ancestors.

In this three parts modular inference, it is possible to complete successfully the first, or first and second, inferences without having sufficient evidence or theoretical background to make respectively the second or third stage inferences. It is possible to prove that there was probably some common cause, some common origins, to a group of species, without having sufficient evidence for the next two stages, finding out the causal-informational net that connected species with origins. If there is more evidence, it may be possible to reconstruct that map, still without inferring the character traits of the origins. Under evidential and theoretical constraints, Darwin stopped sometimes at stage one or stage two, proving that similarities between species were not the result of separate causes but the result of a common descent. I then show how Darwin attempted to prefer the distinct single common ancestor hypothesis over the other five possible common cause models. Finally, Darwin was cautious not to overdraw conclusions at stage three about the properties of ancestral species.

I support this interpretation of Darwin's phylogenic inferences of origins by a close reading of his texts. Then, I criticize alternative interpretations of Darwin's method of inference of origins. Finally, I argue that Darwin's inference of origins belongs to a larger class of inferences of common cause tokens that all follow the same three modular stages that are the distinctive mark of the Historical Sciences.

1.1

Darwin attempted to infer from sets of species information about their origins. The relevant common properties or correlations that distinguish the sets are those that tend *to reliably preserve information*. Their common cause should be the source, the *origin*, of their common properties. The selection, the grouping, of species according to their information preserving qualities must be theory laden. The relevant theories are *information-theories* about the transmission of biological information in time. Explicit discussion of biological information emerged only in the middle of the twentieth century, mostly in connection with the transmission of genetic information via DNA (Artmann 2008). Still, though the concept was not utilized by Darwin—at most he used the expression "transmission"—I analyze Darwin's inference of origins in terms of deciphering a message sent by the ancestor specie to the descendent specie via homologies. Though this may appear anachronistic, it is a fruitful interpretation of Darwin's inference that both clarifies his theoretical assumptions and the inferences based on them, and connects Darwin's inference of common causes from their information preserving effects with other such inferences in other historical sciences.

The simple pre-Darwinian theory of heredity that states that "like begets like," (Sober 1999, 264) "the tendency in every part of the organization, which has long existed, to be inherited" (*Origins*, 359) prepares us to expect correlations between traits of ancestors and their descendents. If so, shared properties could indicate common ancestry. But not all shared properties of species preserve information about ancestry. It is necessary to add to the "like begets like" theory information theories about which similar traits are likely to preserve information about ancestry and which are not. Darwin's theory of evolution is such an information theory about the transmission, selection, and reliability of biological information. This theory helps to identify unreliable information signals about ancestry, homoplasies, because according to Darwinian evolution, it is highly likely that unrelated species that develop useful modifications independently would preserve them and over time would have converging traits without common ancestry. Darwin's theory of evolution, as an information transmission theory, would predict that biological traits that give their bearers great advantages or disadvantages are likely to change relatively quickly in biological time. Traits that give their bearers reproductive advantage will spread quickly in a population. Traits that are disadvantageous will be selected out of the population. Both cannot serve then as reliable information signals for inferring ancestry. These traits are "noise," they can appear and spread quickly, so they do not bear a signal from deep within history. To identify homoplasies as such, Darwin had to make some assumptions about the environments that could lead to convergence of traits. For example, hot climate may lead to homoplastic loss of fur in unrelated animals.

The reliable information bearing signals that are most likely to preserve information about their origins are neutral traits. They do not affect significantly the survival and reproduction chances of their holders, and can reflect the traits of common ancestors. Darwin called them "homologies," "unimportant," "trivial," or "rudimentary." "Homology" is a concept with a long, inconsistent and even confused history (Panchen 1992, 85-108; Amundson 2005, 82-87). But in the context of Darwin's inference of origins, homologies are the traits that are most likely to preserve information about their origins. Common characters that indicate common ancestry "would probably be of an unimportant nature, for the presence of all important characters will be governed by natural selection" (*Origins*, 138). For example, the bluish color of pigeons may not be a reliable indicator of ancestry, but the number of blue markings is indicative of common ancestry:

> [W]e choose those characters which, as far as we can judge, are the least likely to have been modified in relation to the conditions of life to which each species has been recently exposed. Rudimentary structures on this view are as good as, or even somewhat better than, other parts of the organization. We care not how trifling a character may be [...] if it prevail throughout many and different species, especially those having very different habits of life, it assumes high value; for we can account for presence in so many forms with such different habits, only by its inheritance from a common parent. We may err in this respect in regard to single points of structure, but when several characters, let them be ever so trifling, occur together throughout a large group of beings having different habits, we may feel almost sure, on the theory of descent, that these characters have been inherited from a common ancestor. (*Origins*, 337)

I wish to stress that evolutionary theory, in the context of the inference of origins, is an *information transmission theory*. As it so happened, evolutionary theory in biology is also about the evolution of the system of life, not just about the information that system transmitted about its evolution. The same holds true for evolutionary theories in Comparative Historical Linguistics. The theories about the transmission of information about ancient languages are also the theories about the historical evolution of these languages. But this coincidence of theories about the evolution of a system and the evolution of information about its history is not universal. For example, in inferring past states of human society, theories about the evolution of information about human history through documents, diaries, memoirs, traditions and so on are distinct of theories about the evolution of society itself. Historians specialize in the first types of theories, about the transmission of information about the evolution of society; social scientists attempt to theorize the evolution of society itself.

The first stage of Darwin's inference of origins (and Owen's inference of archetypes cf. Ospovat 1981, 146-148; Amundson 2005, 96-98) compares the likelihood of several homologies, "rudimentary" or "trivial" traits of species, given that they preserve information about *some common cause origin* they all share whose *properties are not specified* with the likelihood that the homologous correlations resulted from separate causes. In terms of Bayesian comparison of likelihoods Darwin compared:

$$\frac{\text{Probability (a set of homologous species} \mid \text{some common cause)} \times \text{Probability (common cause)}}{\text{Probability (a set of homologous species} \mid \text{no common cause)} \times \text{Probability (no common cause)}}$$

If $E_1, E_2, ..., E_n$ represent units of evidence, species that share certain homologies; C stand for some common cause origin of this group of species; $S_1, S_2, ..., S_n$, represent separate causes, and B represents background knowledge, the upper part represents the likelihood of the shared homologies, given some common cause; the lower their likelihood, given separate causes. The ratio of the likelihoods determines the choice of explanatory hypothesis between some common cause origins and separate origins:

$$\frac{\{[Pr(E_1 \mid C) \times Pr(C \mid B)] \times [Pr(E_2 \mid C) \times Pr(C \mid B)] \times ... \times [Pr(E_n \mid C) \times Pr(C \mid B)]\}}{\{[Pr(E_1 \mid S_1) \times Pr(S_1 \mid B)] \times [Pr(E_2 \mid S_2) \times Pr(S_2 \mid B)] \times ... \times [Pr(E_n \mid S_n) \times Pr(S_n \mid B)]\}}$$

Darwin assessed the likelihoods of sets of species that share certain traits *given separate causes* by considering the evolutionary advantages of the shared traits. The more evolutionarily advantageous, conducive to survival and reproduction, the more likely are the similarities given separate causes:

> I am inclined to believe that in nearly the same way as two men have sometimes independently hit on the very same invention, so natural selection, working for the good of each being and taking advantage of analogous variations, has some-

times modified in very nearly the same manner two parts in two organic beings, which owe but little to their structure in common to inheritance from the same ancestor. (*Origins*, 162)

Conversely, shared properties that have no evolutionary value are unlikely given separate causes. Darwin repeatedly recognized the significance of "rudimentary, atrophied, or aborted organs—organs or parts [...] bearing the stamp of inutility" (*Origins*, 355, cf. 330) such as the eyes of blind creatures in caves or the wings of birds that cannot fly for phylogenic inference.

Note how Darwin inferred that the transverse bars on the legs of some asses and horses, like those of the zebra, indicate common ancestry by dumping the likelihoods of this similarity of given separate origins:

> He who believes that each equine species was independently created, will, I suppose, assert that each species has been created with a tendency to vary, both under nature and under domestication, in this particular manner, so as often to become stripped like other species of the genus; and that each has been created with a strong tendency, when crossed with species inhabiting distant quarters of the world, to produce hybrids resembling in their stripes, not their own parents, but other species of the genus. To admit this view is, as it seems to me, to reject a real for an unreal, or at least for an unknown, cause. (*Origins*, 142)

Likewise, the morphological structures of the webbed feet of birds and the bones in the extremities of mammals are highly unlikely given separate causes:

> [W]e can hardly believe that the webbed feet of the upland goose or of the frigate-bird are of special use to these birds; we cannot believe that the same bones in the arm of the monkey, in the fore-leg of the horse, in the wing of the bat, and in the flipper of the seal, are of special use to these animals. We may safely attribute these structures to inheritance. (*Origins*, 166)

In *The Descent of Man*, Darwin distinguished three types of homological evidence: Morphology (bodily structure), ontogeny (embryonic development), and rudiments. Again, he compared likelihoods of homologous patterns given some common cause and given separate causes:

> The homological construction of the whole frame in the members of the same class is intelligible, if we admit their descent from a common progenitor, together with their subsequent adaptation to diversified conditions. On any other view, the similarity of pattern between the hand of a man or monkey, the foot of a horse, the flipper of a seal, the wing of a bat, &c. is utterly inexplicable. (*Descent*, 42)

The similar morphologies of man, apes and other mammals and the similar reactions of men and apes to substances and susceptibilities to diseases and parasites (*Descent*, 22-25) are more likely given common causes than separate causes.

Similarities in early embryonic stages of development between human and other mammals also have different likelihoods given common or separate causes (*Descent*, 25-28). For the ontogenic similarities to be more likely given separate causes, there would have to be only one way for life to emerge and develop. Independent life forms would have no other option but to repeat the same stages of embryonic development. Otherwise, the similarities are the result of common ancestry. "No other explanation has ever been given of the marvellous fact that the embryos of a man, dog, seal, bat, reptile, &c., can at first hardly be distinguished from each other" (*Descent*, 42). Darwin inferred from the correspondences between stages of embryonic development of different species that there is a general tendency of embryonic development to correspond with stages of historical evolutionary development, that ontogeny must recapitulate phylogeny. As an information theory, this theory is distinct of the theory of the evolution of the system it signals about. Ontogeny acts as information signals transmitted by the evolution of species in history, through an unclear causal-informational chain. Darwin's historical evolution-embryonic development correspondence theory operates then to extract nested information about ancestry from descriptions of embryonic development.

Darwin devoted most of the discussion of inference of the descent of man to rudiments, though he claimed they are not more important than the other two sorts of homological evidence (*Descent*, 28). Rudimentary organs are

> absolutely useless, such as the mammae of male quadrupeds, or the incisor teeth of ruminants which never cut through the gums; or they are of such slight service to their present possessors, that we can hardly suppose that they were developed under the conditions which now exist. Organs in this later state are not strictly rudimentary, but they are tending in this direction. [...] Rudimentary organs are eminently variable, [...] as they are useless, or nearly useless, and consequently are no longer subjected to natural selection. (*Descent*, 28-29)

For example, several human muscles, the shape of the human ear, the human sense of smell, body hair, and tailbone. The opposite of rudiments were for Darwin *nascent* organs that "though not fully developed, are of high service to their possessors, and are capable of further development" (*Descent*, 28).

Darwin argued that the homologous traits of man and other mammals are highly unlikely given separate causes, and so the common cause hypothesis wins by default:

> If the origins of man had been wholly different from that of all other animals, these various appearances would be mere empty deceptions; but such an admission is incredible. These appearances, on the other hand, are intelligible, at least to a large extent, if man is the co-descendant with other mammals of some unknown and lower form. (*Descent*, 172-173)

Darwin's Inference of Origins

The common cause hypothesis is chosen because of the abysmally low likelihood of homologies given separate cause.

Darwin described the common cause of man and other mammals as "some unknown and lower form." Clearly, he did not want to speculate about its properties, though he did insert without methodological warrant the assumption that the creature was "lower" in some sense, at least lower on the tree of life.

1.2

If some common cause is more probable than separate causes, the next stage of inference is to attempt to determine a causal-informational map that connects all the members of the homologous set. Five alternative types of causal nets may connect the species on a causal-informational net:

(1) *A single ancestral common cause: a distinct common ancestor specie is the common cause of all the descendant species in the set.* The modeling of the history of the transmission of information would be tree-like and be composed of many Y or Ψ like intersections. For example, Darwin believed that it is likely that all life is connected in a single tree with a single origin. "I should infer from analogy that probably all the organic beings which have ever lived on this earth have descended from some one primordial form, into which life was first breathed" (*Origins*, 380).

(2) *Multiple ancestral common causes.* All the species in the set are the hybrid descendants of the same distinct set of ancestral species that interbred and caused the present distribution of common traits among their descendants. The modeling of the history of the transmission of information would be bush-like with W like intersections.

(3) *The common cause may be a member of the set of homologous species:* a species may have several descendant species without going extinct itself. It may then be grouped initially with its descendants, only to be identified later as their ancestor. The modeling of the history of the transmission of information would have K like intersections.

(4) *All the species may mutually cause each other.* All the species may be hybrids. The modeling of the history of the transmis-sion of information would look like a web composed of many H like intersections where information is transmitted between all the units.

(5) *Complex combinations of types 1 or 2, with types 3 or 4.* The set of species may have had either a distinct common ancestor or ancestors, and later either one or more of their descendants interbred with the

others. The phylogenic model would include A like (or upside down A like) intersections.

The discovery of intermediary forms, usually fossils, has assisted in the determination of causal nets since Darwin. But the fossil record was still poor while Darwin was writing. Darwin confronted then difficulties in discriminating between these five possible causal informational model nets ("Y," "W," "K," "H," "A") after he had proved that the evidence is more likely given some common cause than given separate causes. In the case of domestic animals, Darwin just did not know if they were the descendants of a single "wild" species, a "Y" net, or of interbreeding among several ancestral species, a "W" net:

> When we attempt to estimate the amount of structural difference between the domestic races of the same species, we are soon involved in doubt, from not knowing whether they have descended from one or several parent-species. This point, if it could be cleared up, would be interesting; if, for instance, it could be shown that the greyhound, bloodhound, terrier, spaniel, and bull-dog, which we all know propagate their kind so truly, were the offspring of any single species.... I do not believe [...] that all our dogs have descended from any one wild species; but, in the case of some other domestic races, there is presumptive, or even strong, evidence in favour of this view. [...] In the case of most of our anciently domesticated animals and plants, I do not think it is possible to come to any definite conclusion, whether they have descended from one or several species. (*Origins*, 24-25)

Darwin considered several species of domesticated animals concluding that dogs had probably several ancestors (the second "W" model of common cause) while horses, poultry, ducks and rabbits, probably had distinct common ancestors (the first "Y" model). But he could not offer much more than intuition to support his opinion: "most of our domestic animals have descended from two or more aboriginal species must either at first have produced quite fertile hybrids, or the hybrids must have become in subsequent generations quite fertile under domestication. The later alternative seems to me the most probable, and I am inclined to believe in its truth, although it rests on no direct evidence" (*Origins*, 207-208).

The classical tree-like genealogical model in linguistics and evolutionary biology is composed of the first Y types of causal links, a single common ancestor of several species (Wilkins 2009). Interbreeding between existing species challenged the "tree" Y model of descent of distinct species. Darwin's main argument against the prior probability of hybrids or inter-breeding was the alleged sterility of the products of hybrid domestic animals like mules, and of wild species (*Origins*, 363). Sterility or "repugnance to intermarriage" was at the time one of the main arguments against the possibility of interbreeding between species and Darwin adopted it since the very inception of the development of his ideas. The early Darwin thought that hybrids revert to the properties of one of their ancestors, i.e. cease being hybrids and are then indistinguishable from species that have a single common cause, or are infertile (Ospovat 1981, 44-47).

Darwin's second argument against hybrids was based on the assumption that traits are immutable. If so, hybridization cannot account for intermediate forms, only for mixtures of traits. "...the possibility of making distinct races by crossbreeding has been greatly exaggerated. There can be no doubt that a race may be modified by occasional crosses, if aided by the careful selection of those individual mongrels, which present any desired character" (*Origins*, 27). But that race would not have intermediate properties between the two crossed breeds. The existing variety of domestic animals is unlikely given interbreeding that would have created a combination of immutable properties, not intermediary properties. Since the evidence is unlikely given hybridization, there must have been a common ancestor species that through mutations and natural selection led to the present diversity of species. Inconsistently, Darwin argued that the colors of animals cannot serve as rudiments because they can emerge as a result of crossing species with different colors (*Origins*, 138).

Darwin argued specifically against the third K type of common cause that one of the species that share the homologous traits caused the others in the past when they mutated and then evolved separately from the ancestral parent species that continued unchanged. Darwin did not exclude the possibility, but claimed that all else being equal, the prior probability of one species surviving unchanged while having descendent species that evolved quickly, when both compete with each other over similar resources is low:

> It is just possible by my theory, that one of two living forms might have descended from the other; for instance, a horse from a tapir; and in this case direct intermediate links will have existed between them. But such a case would imply that one form had remained for a very long period unaltered, whilst its descendants had undergone a vast amount of change; and the principle of competition between organism and organism, between child and parent, will render this a very rare event; for in all cases the new and improved forms of life will tend to supplant the old and unimproved forms. (*Origins*, 228)

Darwin used a complimentary argument to reduce the prior probability of the fifth possible common cause, A-like, causal-informational network, a combination of multiple origins and later hybridization:

> I am aware that Colonel Hamilton Smith... believes that the several breeds of the horse have descended from several aboriginal species—one of which, the dun, was striped; and that the above-described appearances are all due to ancient crosses with the dun stock. (*Origins*, 140)

Darwin acknowledged that several ur-horses may have caused the present variety of horses including the dun, but he thought it was not probable that the dun then affected all the other races of horses. He doubted the likelihood of the wide distribution of the horse family given the dun as a cause. The wide geographical dispersion of horses excludes frequent crossings with the geographically confined dun. The genealogical causal chains that lead to the present breeds of horses could not have intersected with the dun for reasons of geography. This drops the likelihood of the present geographical distribution of the horse family given a K, H, or A like genealogical models.

In Bayesian terms, Darwin used two arguments to lower the prior probabilities of hybrids. First the argument from sterility that should lower the prior probabilities of all four possible hybrid nets (the W, K, H, and A models), and second, the argument from competition between close species over the same natural resources that should lead to the extinction of the less fit species. This argument should lower the probabilities only of the co-existing ancestor-descendent nets (K. H & A). Then, he used two arguments to lower the likelihoods of homologous groups of species given a process of interbreeding. First, he argued that if the groups include "intermediary forms" they are highly unlikely given interbreeding because properties are fixed and so combining them in interbreeding would lead to mixtures of properties, not to intermediary properties. This argument is against all forms of hybrid nets (W, K, H, and A). Second, a broad geographical distribution of a homologous group is unlikely given a geographical concentration of a possible ancestor species that should interbreed with all the member species in the group. This argument is useful only against the co-existing hybrid types, the K, H, and A like nets. An ancient event of interbreeding, a W like net, can happen in a geographically limited area and the resulting species may migrate later. If the choice between the five possible causal-informational nets of common causation is taken, Bayesian-Sober-style, by multiplying priors by likelihoods and comparing the results, the hybrid networks are going to lose then to the Y like single common ancestry net.

In hindsight, Mendel's experiments with hybrids questioned the assumption of the immutability of traits which formed the basis for Darwin's low estimate of the likelihood of the distribution of traits across species given hybrids. What we know today through DNA analysis about the possibilities of transfer of genetic information from unrelated ancestors to each other, of course weakened very much the argument for low priors of hybrids from sterility. The arguments from competition between related species and the unlikelihood of wide geographical distribution against the co-existence (or at least prolonged co-existence) of ancestor and descendent species may have survived better.

While we have more evidence and better theories than Darwin did, Darwin's inferences were rational given what he had to work with. If we consider the context of reception of Darwin's arguments, as well as the context of his discoveries, there are three different biases that have influenced different scientists in different contexts to prefer the first Y like causal model and exclude the alternative W, K, H, and A hybrid models. First, there is a cognitive elegance bias. It is inferentially and mathematically simpler, neater, and more elegant, to just assume the Markov conditions, to assume independence between units of evidence, that causal and informational chains move only vertically in one direction from ancestors to descendants and not horizontally, between "siblings," even if there is no evidence for it. The "Markovian assumptions are not *a priori* true, but they are entirely standard in causal modeling across the sciences" (Sober 1999, 269). We find this bias in favor of simply assuming rather than proving the causal and informational independence of a group that receives a common cause explanation in Reichenbach's (1956) original philosophical formulation of the problem, though there are reasons to believe that the world is "incestuous" where causal "brothers and sisters" affect each other (Tucker 2007). A reticulate model of phylogenic evolution would have required abandoning the tree model that Darwin adopted and the introduction of a more complex model (Panchen 1992, 58-59).

A second group of biases are political, nationalist and later racist. Nationalists and racists would like to believe that races, nations, and their languages are "pure." They can have common origins, but they do not mix. Darwin's assumption of the sterility of hybrids would have rung familiar from this perspective. It is interesting to note that since this "sterility" criterion for separation of fixed species is obviously not satisfied among the various branches of humanity, the Nazis imposed artificial sterilization on people they considered "hybrids," *mishlingen*, to create artificially the elusive sterility that should have proved the separation between the races of mans.

Thirdly, a progressive bias that considers history (of life, the universe, human society and so on) advancing in the direction of betterment or complexity or intelligence or anything else that is capable of being graded, would favor of progressive Y or at least W like models because they show change, if not progress, and would exclude K, H, and even A like models because they are at least partly continuous and static, non-progressive. Rather than demonstrate historical change through the extinction of common causes such as the origins of species (or languages and so on), they prefer mixing existing or continuous species and traits.

Though Darwin did not display, at least explicitly, the cognitive elegance and political purity biases, he did rely explicitly in some places on the progressive bias against K, H and A like models: for example, the similarities between humans and Old World monkeys that New World Monkeys do not share, indicate that they have some common cause (*Descent*, 180). From the greater homologous similarities with the anthropomorphous apes such as the absence of a tail and callosities and the morphology of man "we may infer that some ancient member of the anthropomorphous sub-group gave birth to man"(*Descent*, 181). This proves the higher likelihood of the set of man and apes given some common cause. Still, Darwin wished to argue further that there was a distinct single ancestor (a Y model) by eliminating at least the third (K) and fourth (H) models that would have suggested that a living ape species is the ancestor of the human race: "...we must not fall into the error of supposing that the early progenitor of the whole Simian stock, including man, was identical with, or even closely resembled, any existing ape or monkey" (*Descent*, 182). Interestingly, Darwin did not consider that man may be the ancestor of apes. But more significantly, he does not propose a particular reason why there must have been a single ancestor of man and apes rather than the other four possible causal informational nets that may connect man and ape.

1.3

Once the most probable causal-information net is chosen, it is possible to compare the probabilities of alternative concrete and detailed common cause hypotheses that specify the traits of the origins. Assessing the probabilities of competing origins hypotheses, requires according to Darwin the examination of the causal-informational chains that transmit information from common cause to species, the fossil record. If the fossil record is missing or incomplete, it may be impossible to accomplish this third stage: "...we should be unable to recognize the parent-form of any two or more species, even if we closely compared the structure of the parent with that of its modified descendants, unless at the same time we had a nearly perfect chain of the intermediate links." (Origins, 227–228)

Rudiments preserve information not just about the existence of some common origin, but also about some of its traits. For example,

> for the progenitor of the upland goose and of the frigate-bird webbed feet no doubt were as useful as they now are to the most aquatic of existing birds. [...] [T]he progenitor of the seal had not a flipper, but a foot with five toes fitted for walking or grasping. (*Origins*, 166–167)

Darwin was usually careful not to overdraw conclusions to the properties of progenitor species when evidence was scarce. Rudiments, by definition, are not of important traits that affected the survival and reproduction of species, so they are hardly sufficient evidence for a comprehensive description of a concrete common cause.

In addition to rudiments, Darwin's inference of the properties of origins relied on morphology, similarities in structure, and ontogeny. The morphology of the body tends to change more slowly, preserve information more reliably, than other aspects of animal physiology, just as the grammatical structures of languages changes more slowly, preserve information about their origins more reliably, than parts of the vocabulary.

On the basis of the three theories about the preservation of information (rudiments, morphology, and ontogeny as recapitulating phylogeny) Darwin was able to infer some of the properties of the ancestors of man: The progenitor of man was covered with hair, had pointed ears capable of movement, a tail, and more muscles than today.

> The great artery and nerve of the humerus ran through a supra-condyloid foramen. The intestine gave forth a much larger diverticulum or caecum than that now existing. The foot was then prehensile, judging from the condition of the great toe in the foetus; and our progenitors, no doubt, were arboreal in their habits, and frequented some warm, forest-clad land. The males had great canine teeth, which served them as formidable weapons. [...]
>
> At a much earlier period the uterus was double; the excreta were voided through a cloaca; and the eye was protected by a third eyelid or nictitating membrane. [...]
>
> At still earlier period the progenitors of man must have been acquatic in their habits; for morphology plainly tells us that our lungs consist of a modified swim bladder, which once served as a float. The clefs on the neck in the embryo of man show where the branchiae once existed. In the lunar or weekly recurrent periods of some of our functions we apparently still retain traces of our primordial birthplace, a shore washed by the tides. At about this same early period the true kidneys were replaced by the corpora wolffiana. The heart existed as a simple pulsating vessel; and the chorda dorsalis took the place of a vertebral column. (*Descent*, 188)

Darwin was cautious about inferring the geographic origin of the progenitor of man. The exclusive presence of the nearest relatives of man, the gorilla and chimpanzee, in Africa indicated in his opinion the probability that man may have originated there. The loss of hair is an indication of a warm climate (*Descent*, 182-183). He predicted correctly that the absence of fossil record for intermediary forms that could have settled the matter reflected the absence of geological excavations in Africa (*Descent*, 184). But since an anthropomorphous ape lived in Europe, and since there must have been sufficient time for migrations, the African origin of man was just more probable than alternative geographical origins. Darwin had no basis for estimating the rate of change among the simian family and therefore could not estimate how long ago humans may have originated.

Another theory that Darwin used in this stage is of "reversion." His evidence for reversion consisted of similarity or correlation not between the common properties of two or more species, but of properties that appear only rarely in one species but are common in the other species. Darwin called the rarer traits "reversions," and explained them as traits that were common in an ancestor species but are far rarer in a descendant species. "The most probable hypothesis to account for the reappearance of very ancient characteristics, is—that there is a *tendency* in the young of each successive generation to produce the long-lost character, and that this tendency, from unknown causes, sometimes prevails" (*Origins*, 141). Darwin used this theory to infer a trait of the origin of the ass, the zebra, etc.

> I venture confidently to look back thousands on thousands of generations, and I see an animal striped like a zebra, but perhaps otherwise very differently constructed, the common parent of our domestic horse, whether or not it be descended from one or more wild stocks, of the ass, the hemionus, quagga, and zebra. (*Origins*, 142)

The very concept of "reversion" presupposes a progressive evolutionary theory of the history of life according to which better adapted species replace less advanced species that become extinct. Without such a theory, there can be no "*reversion*" of some of the properties of a later descendant to those of its earlier ancestor. If accepted, this theory and some unique properties of some individuals in a species may infer some of the characteristics of an ancestor species after its existence and place on the causal net had been established. Darwin seems to rely on progressive teleology, assuming the direction from less to more complex or intelligent. The reversed traits are then of the simpler or less intelligent (*Descent*, 54-62). Otherwise, an exceptionally intelligent ape (Darwin mentioned one in *Descent* in the context of attempting to blur the perceived gap between man and ape) would have been considered evidence for the descent of apes from man. Reversion then is an auxiliary theory that appears in the third stage of inference of origins after the causal-informational net is determined and after the identity of descendants and ancestors is determined.

2

This chapter has advocated a three stage Bayesian analysis of the inference of origins. There have been alternative accounts. Sober (1988, 95) and Forster (1988) interpreted the comparison of likelihoods of a group of species that share some characteristics given common cause and separate causes as between the best common cause hypothesis that specifies the properties of the hypothetical common cause and the best separate causes hypothesis that likewise specifies the properties of the separate causes. There would be then two stages in the inference that a group of species had an exclusive common ancestor:

(1) Two "internal" comparisons among particular common cause origins hypotheses and among particular separate causes hypotheses over which hypothesis makes the group of species most likely.

(2) A final match between the respective "champion" hypotheses of the above semi-final "tournaments."

From a purely logical point of view, there is nothing wrong with this model of inference, if there are sufficient theories and evidence to generate the results. However, Darwin (as well as the founders of historical comparative linguistics) could not conduct comparisons of the likelihoods of homological groups such as that of men and apes given competing concrete hypotheses about their common or separate causes because they did not have such hypotheses. Even if they had, they would not have possessed sufficient evidence to discriminate between them. Likewise, while Rask and Bopp could infer that the homologies of the languages we came to know as the Indo-European languages had a common cause, they could not compare hypotheses about the properties of the hypothetical common cause proto-Indo-European language or about the hypothetical properties of separate independent languages that could have given rise to the present "Indo-European" languages. They did not have sufficient evidence for such a task.

Later, Sober (1999, 259) proposed as an alternative, the comparison of the likelihoods of the group of species given *all* the particular common cause hypotheses, multiplied by their priors, and given *all* the separate causes hypotheses multiplied by their priors. Again, there is nothing wrong with this model from a purely logical or mathematical perspective. But Darwin did not have a list of concrete common cause or separate cause hypotheses whose priors multiplied by likelihoods he could add up, and even if he had, he did not possess sufficient evidence and theoretical background to assign to these hypotheses prior probabilities or compute the likelihoods of the set of species given the sum of these concrete hypotheses.

Sober's (1999) article is entitled "Modus Darwin." It explicitly attempts to reconstruct Darwin's process of phylogenic inference. Yet, it assumes uniform rates of species mutation, which Darwin did not assume. It does not assume the distinction between homologies and homoplasies, which as we notices above is crucial for Darwin's inference of common causes in biology. Instead, it adopts the phenetic as opposed to cladistic (cf. Haber 2009) assumption that all similarities between species are of equal value and there is no independent ground for distinguishing homologies.

Cleland (2009, 57-58) suggested that some inferences of common causes in natural history start with tentative concrete hypotheses about the properties of the common causes, such as ancestral species in evolutionary biology.

In that case, there is no modular inference of phylogenic origins. Phylogenic hypotheses, just like scientific hypotheses in general, are tested individually against each other. Scientists look for a "smoking gun" that can prefer one hypothesis to another. In her opinion, background beliefs determine the prior probabilities of these concrete hypotheses, as well as the classification of the evidence as homologies or homoplasies. Cleland is right about the importance of priors and that the classifications of homologies and homoplasies are theory laden. I argued earlier that information transmission theories play a crucial role in determining these theory laden classifications. But I think it is obvious that Darwin did not possess or claim to possess the kind of concrete hypotheses about origins that Cleland would presume that he must have had. True, the same homologies, rudiments, morphological similarities and similar embryonic stages that participate in inferring that some common cause is more likely than separate causes are also the evidential basis for Darwin's inference of the properties of the ancestral species in the last stage. But, these characterizations are far from comprising together a description of the common cause that can resemble detailed descriptions of living species. Such descriptions of extinct species would become possible only with the addition of fossil evidence. Most significantly, in some cases, Darwin made it abundantly clear that he knew nothing of the properties of the origins, not even if there were more than one common ancestor, of some domestic animals for example.

3

Several historical sciences are concerned with inferring common causes or origins: contemporary phylogeny and evolutionary biology infer the origins of species from information preserving similarities between species, DNAs and fossils; Comparative Historical Linguistics infers the origins of languages from information preserving aspects of existing languages and theories about the mutation and preservation of languages in time. Darwin compared species to languages and phylogenic inference to the inference of ancestral languages (*Origins*, 334-335). "Rudimentary organs may be compared with the letters in a word, still retained in the spelling, but become useless in the pronounciation, but which serve as a clue in seeking for its derivation" (*Origins*, 359). The distinction between homologies and homoplasies is just as useful in Historical Linguistics:

> We find in distinct languages striking homologies due to community of descent, and analogies due to a similar process of formation. [...] The frequent presence of rudiments, both in languages and in species, is still more remarkable. [...] In the spelling also of words, letters often remain as rudiments of ancient forms of pronounciation. (*Descent*, 113)

Information about common origins may be lost through an evolutionary process. Species and languages can spread and exterminate others. Processes of natural selection may affect the reliability of information nested in present languages about their ancestors, just as in present species. Once linguists demonstrate that the homologies between languages are more likely given some common cause than given separate causes, the next stage is to find out the causal-information net that connects them since "distinct languages may be crossed or blended together" (*Descent*, 113).

Darwin's inference of origins is a particular case of a general model of inference of common cause tokens from multiple sources of evidence that preserve similar information about them (Tucker 2007, 2004). As Hull (1992) recognized, one of the basic tasks of the historian is to distinguish patterns that result from natural regularities such as homoplasies, which have separate causes (often different tokens of the same type), from those that have common cause tokens, such as homologies. Archaeology infers the common causes of present material remains; and Cosmology infers the origins of the universe. These are the Historical Sciences, sciences that attempt to infer rigorously descriptions of past events and processes from their information preserving effects. By contrast, other sciences that we may call the Theoretical or Experimental Sciences are not interested in any particular *token* event, but in *types* of events: Physics is interested in the atom, not in this or that atom at a particular space and time; Biology is interested in the cell, or in types of cells, not in this or that token cell; Economics is interested in modeling recessions, not in this recession; and Generative Linguistics studies "Language" not any particular language that existed in a particular time and was spoken by a particular group of people. The theoretical Sciences are interested in inferring regularities between types from replicated experiments.

If we revisit the Neo-Kantian question about the distinction between the sciences that Windelband and Rickert raised, we have discovered a new epistemically and methodologically founded alternative. Two types of inference, of common cause tokens and common cause types distinguish the historical sciences from the theoretical sciences. Darwin presented a paradigmatic case of the methodology of the historical sciences, just like Rask, Bopp, and Ranke.

Darwin's phylogenic inferences follow three modular stages. First, he used the theory of evolution by variation and natural selection as an information theory to distinguish homologies including morphological features from homoplasies. The theory that ontogeny recapitulates phylogeny is distinct of the theory of evolution but is also useful for inferring that a set of species had some common cause. Then, in the second stage he used again the theory of evolution and the assumption of sterility of hybrids to lower the prior probabilities of hybrid hypotheses. He used the fixity of traits and the wide geographical dispersal of some species to reduce the likelihood of the present distribution of traits given hybridity. Having established single ancestry, Darwin used the previously identified rudiments and morphological homologies together with the theory that ontogeny recapitulates phylogeny to try and infer some of the characteristics of ancestor species. He recognized that in some cases, there was insufficient evidence for this task. The theory of evolution plays here a duel role. On the one hand, it is one of the theories that assist Darwin in his phylogenic inferences. On the other hand, the results of Darwin's phylogenic inferences supported the confirmation of his theory of evolution. There is a measure of circularity here. But this is not vicious circularity, since the theory of evolution "bootstraps" itself in Clark Glymour's (1980) sense.

Abbreviations

Descent: Ch. Darwin. *The Descent of Man, and Selection in Relation to Sex*. London: Penguin, 2004.
Origins: Ch. Darwin, *The Origin of Species By Means of Natural Selection*. New York: Barnes & Noble Classics, 2004.

References

Amundson, R. (2005). *The Changing Role of the Embryo in Evolutionary Thought*. Cambridge: Cambridge University Press.
Artmann, S. (2008). Biological Information. In *A Companion to the Philosophy of Biology*. S. Sarkar and A. Plutynski (eds.). Malden, Mass.: Blackwell, 22-39.
Cleland, C. (2009). Philosophical Issues in Natural History and Its Historiography. In Tucker (2009), 44-62.
Forster, M. R (1988). Sober's Principle of Common Cause and the Problem of Comparing Incomplete Hypotheses. *Philosophy of Science*, 55: 538-559.
Glymour, C. (1980). *Theory and Evidence*. Princeton: Princeton University Press.
Huber, M. (2009). Phylogenic Inference. In A. Tucker (ed.). *A Companion to the Philosophy of History and Historiography*. Malden, Mass.: Wiley-Blackwell, 231-242.
Hull, D. (1992). The Particular-Circumstance Model of Scientific Explanation. In *History and Evolution*. M. H. Nitecki and D. V. Nitecki (eds.). Albany, N.Y.: State University of New York Press, 69-80.
Ospovat, D. (1981). *The Development of Darwin's Theory: Natural History, Natural Theology, and Natural Selection, 1838-1859*. Cambridge: Cambridge University Press.
Panchen, A. L., (1992) *Classification, Evolution, and the Nature of Biology*, Cambridge: Cambridge University press.

Sober, E. (1988). *Reconstructing the Past: Parsimony, Evolution, and Inference.* Cambridge, Mass.: The MIT Press.
_____ (1999). Modus Darwin. *Biology and Philosophy* 14: 253-278.
_____ (2002). Reconstructing the Character Traits of Ancestors: A Likelihood Perspective on Cladistic Parsimony. *The Monist* 85 (1): 156-176.
Tucker, A. (2004). *Our Knowledge of the Past: A Philosophy of Historiography.* New York: Cambridge University Press.
_____ (2007). The Inference of Common Cause Naturalized. In *Causality and Probability in the Sciences.* J. Williamson and F. Russo (eds.). London: College Press, 439-466.
_____ (ed.) (2009). *A Companion to the Philosophy of History and Historiography.* Malden, Mass.: Wiley-Blackwell.
Wilkins, J. S. (2009). Darwin. In Tucker (2009), 404-415.

7

The Scientific Status of Darwinism

Vladimír Havlík

Darwin's theory of natural selection has profoundly influenced not only biology and other scientific disciplines, but also the considerations of the nature of science in general. Whether offering general normative rules with which to bind scientific practice or, alternatively, denying the very existence of a general scientific method, philosophers have always drawn on examples from the history of science—most importantly, the history of exact disciplines such as physics. This was the science which served as the basis for the rules of confirmation and falsification that were supposed to be adhered to across all scientific disciplines. Within such an approach, Darwin's theory appears quite out of place, since it does not satisfy the mathematical and experimental requirements of physical theories. Thus it presents us with a dilemma: in view of the exceptional character of Darwinism, we should *either* rethink the nature of science, *or* deny that Darwinism counts as science. The aim of this chapter is to show that it is not necessary to see Darwinism as a science merely in some minimal—semantic or explicatory—sense, but, on the contrary, that the core of this theory satisfies the strictest criteria of the hypothetico-deductive (HD) model of science. At the same time, Darwin's theory does, to be sure, have its own specificities. These shall also be pointed out in what follows.

At the first approximation, the difficulty in determining the scientific character of Darwinism follows already from the fact that it is hard to capture the content of this theory. While we can distinguish various phases in the development of Darwinism on the basis of important discoveries, it is virtually impossible to express the content of the theory of Darwinism as it is understood today. This has several reasons. For the one thing, there are theoretical disagreements within Darwinism (e.g., alternative views of the mechanism of evolution). For another, there are theoretical conceptions that often start as radical alternatives (e.g., the theory of punctuated equilibria), but get assimilated by the modern evolutionary synthesis later (see Okasha 2000). There are several theoretical approaches whose proponents distance themselves from Darwinism, and it is an open question whether these approaches will be assimilated within the Darwinist paradigm in the future.

It may seem somewhat unjustified to use the term "Darwinism" to refer to the modern evolutionary synthesis, rather than merely to the specific form of Darwin's theory as laid out in the *Origin of Species*. Ernst Mayr, for example, identifies five principles in Darwin's original theory, of which the first two—namely, the fact of evolution and the thesis of common descent—he takes as constituting the first Darwinian revolution, which took place shortly after the publication of *Origin* (see Mayr 2004, chap. 6). The remaining three principles—gradualism, speciation and natural selection—he links to the second Darwinian revolution that took place during the evolutionary synthesis. Moreover, one must emphasise Darwin's original rejection of Lamarck's mechanism of inheritance, and the so-called "neo-Darwinism" of Weissmann and Wallace who rejected Lamarckism as well. Mayr therefore suggests that the Darwinism which has taken root since the evolutionary synthesis be called simply "Darwinism," because it essentially coincides with Darwin's original conception of 1859—except that the Lamarckian idea of inheritance of acquired characteristics, with which Darwin himself toyed, has been completely rejected by contemporary researchers.

When I speak of the scientific character of Darwinism, I do not take this theory as static or complete. Rather, I see it as developing and extending the original core of the aforementioned central principles in unforeseen directions. There is an analogy in physics. Current versions of the theory of relativity are not bound by Einstein's original formulations. Not only the way this theory is formally *expressed*, but its *content* as well, has changed over the years, as new discoveries provided solutions for the theory's basic equations and their applications in other fields of science. But Darwinism is even more complicated than that—or so I shall argue in the rest of this chapter.

So let us start with the assumption that Darwinism is a constantly developing way of thinking linked to a relatively fixed core. What is the structure of this theoretical edifice? One can say that in attempting to clarify Darwin's core theory of natural selection as laid out in the *Origin*, philosophers have already used pretty much everything on offer in the toolbox of contemporary philosophy of science. The philosophical effort to clarify the scientific status of Darwinism is interesting not only at the level of the methodology of the formulation of scientific theories—i.e., at the level of answering questions, such as whether Darwin's approach adheres to the deductive rather than inductive method, or whether it exemplifies several alternative strategies, etc.—but also from the point of view of meta-methodology—i.e., at the level at which one inquires which philosophical methodology should be employed in analysing Darwin's theory. If we set aside the notion that Darwin's theory is metaphysical rather than scientific (Popper 1974, 195), we find a whole range of analyses of the scientific nature of Darwinism. Thus, there are attempts to fit it in the traditional HD model (Ruse 1975;

Ghiselin 1984); to see it as an example of inference to the best explanation (Thagard 1978); of the probabilistic causal theory (Hodge 1977 and 1987); or of the theory of forces (Sober 1984). Yet others have interpreted Darwinism in terms of the semantic theory model (Lloyd 1983; Sintonen 1990); as a prime example of unification within science (Kitcher 1985); as a paradigmatic theory in the Kuhnian sense (Wilson 1992); or as an instance of mechanistic explanation based on the concept of natural selection as a mechanism (Barros 2008). It should be evident that, while some of the philosophical approaches on my list are more opposed to each other, others can be taken as complementary. At any rate, my aim is not a detailed review of these various proposals and their mutual relationships. Rather, I wish to capture the specific nature of the Darwinian science by pointing out some of the key features that emerge from the variety of philosophical approaches to Darwinism.

I believe that the lack of consensus concerning the logical character of Darwin's theory should not be mistaken for a proof that Darwinism fails as a science. From the point of view of meta-methodology, the various philosophical approaches to Darwinism listed in the previous paragraph can be understood as "methodological research programmes" (Lakatos 1970) that apply their models to the logical and historical development of a scientific theory and to its conceptual reconstruction as well. In other words, these different philosophical analyses of Darwinism should be taken as different *perspectives* from which to see a particular scientific theory. These perspectives, of course, have fundamental consequences for how we conceive of science as such. However, with the exception of Popper's views, the other approaches on our list do not challenge the scientific character of Darwinism; while they might otherwise disagree on what exactly its scientific nature consists in, they do agree at least on this much.[1]

Yet an assumption, however widely shared, does not amount to a demonstration. Hence, while I shall not, to repeat, attempt to reconcile the disagreements among various philosophical approaches to Darwinism, I shall draw on some of the features of this theory that have emerged from these debates, and use them in an argument in support of the shared assumption about the scientific nature of the theory. In the process, it will become clearer what the special character of Darwinism, compared to other sciences of nature, consists in.

One of the features of Darwinism often mentioned by different authors is that it can be at least partly axiomatised, and thus conform to the HD model of scientific theory. To be sure, not all the parts of the *Origin* can be easily ren-

---

[1] Moreover, Popper himself eventually gave up his original view that Darwin's theory was untestable and tautological (see Popper 1978, 355). And even at the time when he doubted the testability of Darwinism, he acknowledged its enormous explanatory power and saw it as an instance of a trial and error theory—i.e., a theory fit for achieving the truth (see Popper 1974, 195).

dered in an axiomatic form. In particular, Darwin's reasoning in terms of an analogy between artificial and natural selection resists this expression. Some authors see this as a serious shortcoming in Darwin's theory—they either see the theory as a "mere shot at a hypothetico-deductive model," or conclude that it is basically inductive in its form.

However, Michael Ruse believes that the case for conforming Darwinism to the HD model is not hopeless. This is because Darwin himself meant to propose a theory conformed to this model, his appeals to an analogy between artificial and natural selection notwithstanding (see Ruse 1975, 233). In support of this interpretation, Ruse quotes from Darwin's *Notebooks* as well as from the *Origin*. For example, in his Notebook B, Darwin says:

> Astronomers might formerly have said that God ordered each planet to move in its particular destiny. – In same manner God orders each animal created with certain form in certain country, but how much more simple & sublime powers let attraction act according to certain law such are inevitable consequen[ces]. Let animals be created, then by the fixed laws of generation, such will be their successors. (Darwin 1837-1838, 101)

In the conclusion of the *Origin*, Darwin argues that from such laws, "the production of the higher animals directly follows" (Darwin 1859, 490). It seems, however, that Darwin does not mean to demonstrate here the HD character of his theory, as he tried to present it in *Origin*. Rather, has in mind the law-like character of the evolutionary process itself, i.e., its necessity and generality. They way he managed to express his theory was not, at the moment, his focus.

This view finds support in many other passages from Darwin's works. For example, when comparing his theory with Newtonian physics, Darwin was clearly aware that the latter was the model of a scientific theory and that the laws of motion can be directly tested and confirmed (see Lloyd 1983). By contrast, he took for granted that natural selection cannot be directly tested by evidence: "[...] we can prove that no one species has changed [*i.e.* we cannot prove that a single species has changed]; nor can we prove that the supposed changes are beneficial, which is the groundwork of the theory" (Darwin 1887 [vol. 3], 25). Moreover, in his letters, Darwin does not speak of a *theory*, but of a mere *hypothesis*. His concept of the relationship between a theory and a hypothesis, however, fully depends on the latter's ability to provide an explanation for a variety of different phenomena. "I have always looked at the doctrine of Natural Selection as an hypothesis, which, if it explained several large classes of facts, would deserve to be ranked as a theory of deserving acceptance" (Darwin and Seward 1903 [vol. 1], 140-141). Darwin pointed out the wave theory of light as an example from the domain of physics that shows how an hypothesis can turn into a theory. In many of his letters, he emphasised that ether is also hypothetical and wave motion is merely deduced from the explanation of light, and that therefore "[...] an hypothesis is *developed* into a theory solely by explaining an ample lot of facts" (Darwin 1887 [vol. 2], 286).

Given the form of his argument for natural selection, Darwin faced the criticism that he abandoned the spirit of inductive philosophy (see Darwin and Seward 1903 [vol. 3], 148). Accordingly, he tried to defend the legitimacy and scientific character of his hypothesis. "I should really much like to know why such an hypothesis as the undulation of the ether may be invented, and why I may not invent [...] any hypothesis, such as Natural Selection" (ibid.). Moreover, Darwin's argument draws on a broad class of empirical evidence in support of his hypothesis of natural selection. He says that

> [...] no theory so well explains or connects these several generalizations [...] as the theory, or hypothesis [...] of Natural Selection. Nor has any other satisfactory explanation been ever offered of the almost perfect adaptation of all organic beings to each other, and to their physical conditions of life. (Darwin 1863)

In all these passages, we can appreciate not only Darwin's views about the nature of science and its methodology, but also how the very same passages can be used in support of alternative interpretations. It can be reasonably assumed that Darwin's texts might support all the philosophical interpretations of Darwinism that I listed several paragraphs above. Hence it is very important to distinguish between intent and execution, between Darwin's intention and his actual results. As noted by Ruse, "obviously, there is no logical connection between what a man intends and what he achieves" (Ruse 1975, 233). What should, then, be decisive in formulating a meta-methodological rule with which to evaluate the scientific character of a candidate theory—assuming that it is possible to lay down such rules to begin with? Let's assume that Darwinism is precisely a testable case for the successful formulation of meta-methodological rules for the evaluation of scientific theories. Couldn't we, then, use this very specificity of Darwinism in the opposite sense—i.e., not only to prove this specificity itself, but also to test the capacity and rationality of various meta-methodological concepts and models? For what is decisive in this diversity of opinions concerning the character of Darwin's argument? Is it the role of laws, analogies, forces, semantic content or explanation? And, finally, are not the differences among the individual models of scientific theory too intricate and idealised, so that they cannot be satisfied by any actual scientific theory? I contend that a consideration of the special character of the Darwinian science can help answer at least some of these questions.

In order to demonstrate this specificity, let's first return to the possibility of the axiomatisation of Darwin's theory. It is indeed possible to axiomatise Darwin's original theory by means of mutually independent axioms and their logical consequences, and thus to demonstrate the structure of such a theoretical model. Of course, Darwin never speaks of an axiomatisation of his theory.

He starts from some facts that he established by means of countless observations. He writes:

> In considering the Origin of Species, it is quite conceivable that a naturalist, reflecting on the mutual affinities of organic beings, on their embryological relations, their geographical distribution, geological succession, and other such facts, might come to the conclusion that each species had not been independently created, but had descended, like varieties, from other species. (Darwin 1859, 3)

Darwin's task is, then, to find an explanation of the fact of the organic diversity of species on the basis of given empirical evidence. Is it possible to discover a single cause of this diversity of life, a mechanism behind the origin of species that would do away with the idea of special creation? Darwin eventually arrives at a certain unifying logic connecting the given data—the logic that admits only natural causes.

Despite its familiarity, let's remind ourselves of the key elements of this logic, i.e., of the mechanism of natural selection. Darwin drew on the following facts as evidence in favour of natural selection: (1) the variety of species; (2) the differential fitness of their traits; and (3) the heredity of these traits—i.e., the ability of parents to impart their traits on their descendents. Darwin also looked at the domestication of animals and the cultivation of plants that provided evidence both for the fact of heritable modification, but also for the possibility of accumulation of modifications by means of *selection* of the desired traits (see Darwin 1859, chap. 1). However, the mechanism had to explain not only the perfect forms of a species, but also, more importantly, how individual organisms get adapted to each other and to their environment. In order to explain these things, Darwin adopted two elements of the Malthusian population theory, namely (4) the geometric growth of population, and (5) a limited environment with scarce resources for such a population. Under these constraints, a struggle for existence necessarily followed. Darwin writes:

> As many more individuals of each species are born than can possibly survive; and as, consequently, there is a frequently recurrent struggle for existence, it follows that any being, if it vary however slightly in any manner profitable to itself, under the complex and sometimes varying conditions of life, will have a better chance of surviving, and thus be *naturally selected*. (ibid., 5; emphasis in the original)

For my present purposes, I can leave aside many details of Darwin's empirical evidence. I am concentrating on the logical structure of his argument in favour of the mechanism of natural selection. Following Ruse, we can summarize Darwin's argument as follows (Ruse 1975, 222):

(1) Variability of species.

(2) Differential fitness of traits.

(3) Heritability of these traits.

(4) Reproduction of organisms by geometric growth

(5) Finite space with limited resources.

From these propositions, it logically follows:

(C) Natural selection of the species.

The evolution by means of natural selection is thus a process which necessarily occurs, given the conditions (1)–(5) are satisfied. The axiomatised theoretical structure of the process of evolution is thus generalised, unified and abstracted in a way similar, e.g., to the formulation of the physical law of gravitation. This approach also makes it possible to state the so-called "universal Darwinism," which takes organic evolution to be one of the many instances of a universal evolutionary algorithm (see Dawkins 1976).

I should like add a few critical remarks to the axiomatisation of Darwin's argument as proposed by Ruse. Firstly, in his summary of Darwin's theory, Ruse follows the chapters of the *Origin*. He thus assumes that the form as well as the content of Darwin's book is important for the reformulation of his theory. In other words, Ruse understands the order of exposition of Darwin's theory in the *Origin* as integral to the theory, despite the fact that Darwin himself noted that his book was only a quickly composed and imperfect précis (see Darwin 1859, 2). Therefore I take it that Darwin himself allowed that we should separate the logical content of his theory from its actual presentation, and attempt a reconstruction of the theory in a deductive form. Such a separation of the logical content from its presentation makes it also easier to address the analogies used by Darwin, which many authors reject as inadmissible in a theory that has a deductive form.

Secondly, despite the fact that Darwin uses the term "struggle for existence," it seems unnecessary to include it in the axiomatisation of his theory without any loss of its content. The term "struggle" smuggles into the theory an untestable element, since it misleadingly puts emphasis on the capacities of individual organisms, while the crucial thing for the argument is the limited space and resources. Therefore, I believe that it is better to drop the phrase "struggle for existence."

Thirdly, Ruse does not consider Darwin's argument as strictly speaking deductively valid because, since its premises are not guaranteed to be true. For example, struggle for existence might not occur, if some catastrophe wiped out large numbers of organisms. In other words, axioms (3) and (4) might be satisfied, and yet struggle for existence would not occur. But Ruse thinks it possible to make the argument deductively valid by adding some premises that Darwin considered obviously true but did not mention explicitly (see Ruse 1975, 222). In my view, Ruse's line of reasoning is flawed. The validity of a deductive argument is independent of an actual state of affairs. Accordingly, even if empirical conditions changed radically—e.g., even under the condition of mass extinction—the deductive validity of Darwin's argument wouldn't be affected. A mass extinction must be understood as a result of an external force which is independent of the mechanism of evolution itself, even though it could affect its implementation. It's easy to find analogies in the domain of physics. For example, if a physical body in a gravitational field is deflected from its expected trajectory as a result of its collision with another body, the deductive structure of the law of gravitation is not affected. An accidental mass extinction of organisms is thus analogous to the collision of two physical bodies.

A similar conclusion could be also reached by following the distinction, suggested by E. S. Reed, between *explanation* by means of natural selection and *applicability* of natural selection: "natural selection does not have universal explanatory power, but it does have universal applicability" (Reed 1981, 64). I believe this distinction might help resolve many disputes about the scope of application of natural selection. In virtue of Reed's distinction, we can both claim every species evolved by natural selection, and allow the effect of external factors, such as speciation, genetic drift or punctuated equilibria. Analogously, according to the law of gravitation, every physical body is subject to gravitational force, and yet there are many other intervening factors, such as aerodynamic lift (ibid., 65). This interpretation is confirmed by Darwin's own remarks to the effect that natural selection is the principal mechanism of modification, although it is no means exclusive (Darwin 1859, 6).

However, the logical structure of Darwin's argument offered so far does not express the core of Darwinism that I promised to delimit. As mentioned earlier, Mayr mentions five mutually independent principles (or theories) of Darwinism: the fact of evolution of species as such; common ancestor; gradualism; speciation; and natural selection (Mayr 2004, chap. 6). Of those, the argument summarized by Ruse includes only the principle of selection (i.e., variability), and natural selection occurs in the argument as its conclusion, instead of premise. But the rest of the principles can be regarded as axioms of the theory of Darwinism: they do not contradict the other principles and they are not deductible from them, either. Moreover, Darwin believed that there was ample empirical evidence for all of them.

Darwin suggests the logical structure of his hypothesis in a letter to G. Bentham:

> In fact the belief in Natural Selection must at present be grounded entirely on general considerations. (1) On its being a *vera causa*, from the struggle for existence; and the certain geological fact that species do somehow change. (2) From the analogy of change under domestication by man's selection. (3) And chiefly from this view connecting under an intelligible point of view a host of facts. (Darwin [vol. 3], 25)

However, this letter as well as other sources suggest that Darwin took the analogy between natural and artificial selection to be an important aspect of his theory. However, as noted earlier, analogy, being an instance of inductive reasoning, is strictly speaking inadmissible within a deductive theory. Thus we face a dilemma–either we must give up the deductive character of Darwin's theory, or try to preserve it by eliminating the analogical reasoning. Ruse considers analogy an integral part of Darwin's theory; therefore, he thinks it impossible to remove it. He believes the core of Darwin's theory is built in part deductively, in part analogically, and that it is then applied in explaining the variety of biological phenomena (see Ruse 1975, 240). I agree that the structure of Darwin's theory is much more complex than many commentators assume. And yet, Ruse's inclusion of both deductive and inductive arguments seems to me to be too much of a compromise.

I wonder whether Ruse does not understand the role of analogy in Darwin's theory incorrectly. Darwin clearly introduced an analogy between artificial and natural selection for strategic reasons: he wished to gain support for his discovery of natural selection as the key mechanism of evolutionary change. Ruse himself notes that analogical reasoning in Darwin plays the role of a *support* as well as a *heuristic*. And yet, Ruse insists that this part of Darwin's theory cannot be deductively linked to the rest (ibid., 226).

However, let us consider two facts. First, in writing the *Origin*, Darwin was not after an axiomatisation of a strictly deductive theory. Rather, he meant to make as strong as possible a case for the existence of a true cause (*vera causa*) of the diversity of species and their adaptation to their environment–the case, moreover, that would convince as wide audience, both expert and lay, as possible. Second, in his ideas about the method of science, Darwin was heavily influenced by such nineteenth-century philosophers of science as Herschel and Whewell, who both understood induction as fundamental to scientific reasoning. Owing to these philosophical influences, Darwin understood an appeal to analogy an advantage. Now, should we conclude, with Ruse, that the core of Darwin's theory is built from inductive as well as deductive elements? I believe that either a strategy, chosen for the exposition of a theory, or a philosophical legacy that influenced it, is not necessarily integral to the theory itself. In other words, I am appealing, again, to a distinction between a logical content and a form of a theory. Analogical reasoning is part of the form, not the content. Let's add to the axioms (1) to (5) another:

(6) Under certain circumstances, traits can be intentionally selected.

As a result, artificial selection deductively follows from the theory just like natural selection. Artificial selection is possible if and only if natural selection is possible. Artificial selection is but an application of the laws of natural selection under certain special circumstances. What we are dealing with here is a single process that assumes either intentional or unintentional form. Hence, the accidental fact that Darwin chose to accommodate his audience by presenting natural selection in analogy with artificial selection does not affect the deductive character of his theory.

Moreover, analogy occurs when, for example, the same method is applied in different domains, or when we reason on the basis of facts from one domain to the facts from another domain. Thus, Darwin appeals to an analogy between the undulation of ether and the motion motion air in the transmission of sound:

> The ether is hypothetical, as are its undulations; but as the undulatory hypothesis groups together and explains a multitude of phenomena, it is universally now admitted as the true theory. The undulations in the ether are considered in some degree probable, because sound is produced by undulations in air. So natural selection, I look at as in some degree probable, or possible, because we know what artificial selection can do. (quoted in Sarton 1937, 340)

In the case of ether, the analogy is completely justified. As sound is transmitted by means the waves of a physical medium of one sort—namely, air—so is light supposedly transmitted by means of the waves of a physical medium of a hypothetical sort—namely, ether. There are two different media, or physical domains. However, in the case of an analogy between artificial and natural selection, the condition of the essential difference between domains is not satisfied. What we have here is a single process implemented under different conditions, so that we can't speak of analogy in a strictly inductive sense. Rather than an analogy, we are dealing here with an extension of applicability of certain phenomena. Therefore, we should understand Darwin's analogy in this weak sense. When duly axiomatised, it would also fit into the structure of a deductive theory. Finally, I should like to add that physical and mathematical theories have also been reformulated and axiomatised in order to best exhibit their deductive structure. An initial lack of consistency cannot affect the logical character of a theory—and Darwin's theory most likely in no way differs from others in this respect.

I believe that we need to distinguish between a logically deductive core of a theory from the form of its presentation in the case of laws and initial conditions. The HD model requires that initial conditions are axiomatised and that the statements of a theory assume a lawlike form—i.e., they are universal and general. Ruse thinks that Darwin's theory fits the HD model even less in the case of laws than in that of its logical structure. Despite that, we can still consider some of the statements in Darwin's argument as laws: for example, organisms tend to propagate in a high degree (see Ruse 1975, 223). Yet others lack universal applicability and merely describe particular entities. Although Ruse finds a way to turn such propositions into laws, the end result departs from Darwin's own theory. Accordingly, Ruse concludes that "Darwin's arguments, taken literally, are neither deductive nor solely composed of laws" (ibid. 1975, 224).

I do not accept this verdict. For one thing, Ruse does not clearly distinguish between the putative lawlikeness of a statement about the propagation of organisms, on the one hand, and the putative particularity of a statement about the limits of space that can be inhabited by organisms, on the other. It seems to me that both of these statements make claims about the properties of particular things. Furthermore, we must not confuse the lawlikeness of the core of a theory with its presentation, on the one hand, and the laws with initial conditions, on the other. Darwin did not conform his reasoning to the HD ideal either in the case of the overall logical structure of his theory, or in the present case of lawlike nature of the statements of the theory. Rather, he took as his point of departure obvious empirical evidence, such as the variability of organisms, their tendency to propagate in a high degree, and the limited space and resources. These initial conditions play the role of universal statements. Hypothetical cases, be it the existence of organisms that do not change over time, or organisms that multiply yet do not increase in number, or, again, the worlds with limitless expanses of space do not, in my opinion, contradict the lawlikeness of Darwin's initial conditions. Moreover, no statements that qualify as laws satisfy the requirement of complete universality and generality. Newton's gravitation law applies only under certain non-relativistic conditions. We could say that laws are valid relative to a domain. Empirical statements that describe the initial conditions in Darwin's theory are lawlike within a particular domain—the only domain so far known in which the evolution of organic beings takes place.

However, there are other points of contact between biology and physical theories that suggest that they are scientific in a similar manner. For example, Darwin was able to confirm, but not explain, his axioms. He knew neither the mechanism of heredity, nor the nature of forces that are behind the striving for a preservation and propagation of life, irrespective of available space and resources. We might say that Darwin's evidence was *phenomenalistic*, in the sense that he was able to collect it, but not—given the limitations of his knowledge—to explain it. In the explanation of heredity, it was necessary to dig up finer structures than those of species, organisms, or even cells. Other axioms of Darwin's theory have not been satisfactorily explained to this day, despite a remarkable progress.

Now the situations in which we are capable of collecting evidence, and yet their explanation escapes us, is by no means exceptional in sciences. On the contrary, Elliott Sober shows they are quite common. For example, the correlation between the phases of the Moon and the rise and fall of sea levels had been known long before Newton explained it (see Sober 1984, 22). Yet Sober's example is not quite well chosen, if what we are after is a case analogous with a mismatch between Darwin's evidence and his inability to explain it. In Sober's example, the knowledge of a correlation did not lead to a theoretical construct that could claim the status of a scientific theory. For this reason, we should rather compare Darwin's theory with some well-established scientific theory. Let's consider phenomenological thermodynamics. This is a physical theory that starts from several experimentally confirmed properties and dependencies, axiomatises them, and derives from them certain other properties and relations. Phenomenological thermodynamics neither questions nor explains its initial axioms, such as temperature, pressure or capacity. Instead it searches for lawlike dependencies among such empirical facts at its own level—whatever the ground of these facts at some deeper level might be. Darwin's theory seems similar, at least in its original form. That is, it was originally proposed as a phenomenological theory. Contemporary Darwinism preserves the original phenomenological core, but adds to it a much deeper theorizing of the ultimate mechanisms of evolutionary process.

This is exactly why we should see Darwinism, compared to physical theories, as truly unique. Speaking of the special nature of Darwinism, I do not mean that it is not a deductive theory. Every empirical theory could be reconstructed in a purely deductive form, if we disregard its relations both to empirical reality and to scientific practice. I have argued that Darwin's theory is as deductive as other scientific theories, provided that we abstract from these particular relations. The special nature of Darwinism consists only in the complexity and multiplicity of levels that have been added to the original argument for natural selection over the last 150 years.

It is precisely this multi-level character of Darwin's evolutionary theory which eludes the common concept of scientific theory. The core of Darwinism is covered with layers of detail not only within its original phenomenological domain, but also at many other levels of biological reality—genes, molecules, cell structures, the immune system, organs, organisms and populations, species and higher cladistic structures. Furthermore, evolutionary biology is being enriched by new theoretical approaches to the study of autonomous and complex systems. This is a unique feature of biology. Physics and chemistry are limited to their own domains and do not expand to either higher or lower levels of reality. So while physical sciences enjoy a relative autonomy within their proper domains, Darwinism is due to its multiplicity of levels without a parallel in contemporary science.

In my view, this is exactly the reason why philosophers find it so difficult to pick the standards in terms of which to judge the scientific character of Darwinism. If Darwinism needs to respect the specific features of different levels of reality and their complex mutual relationships, and philosophers who study Darwinism look at different aspects of this complex science, then it is not surprising that they have a hard time reaching consensus. And yet, the HD core of Darwinism has so far survived in an environment of many new and fundamental discoveries and theoretical innovations. Hence it demonstrates a remarkable *fitness* in the reproduction of scientific ideas. The future philosophical analyses of Darwinism should thus take for granted its status as a falsifiable and confirmable scientific theory.

References

Barros, D. B. (2008). Natural Selection as a Mechanism. *Philosophy of Science* 75: 306-322.
Barrett, P. H. et al. (ed.) (1987).*Charles Darwin's Notebooks, 1836-1844. Geology, Transmutation of Species, Metaphysical Enquiries*. Cambridge: Cambridge University Press.
Darwin, C. (1964 [1859]). *On the Origin of Species. A Facsimile of the First Edition*. Introd. by E. Mayr. Cambridge, Mass.: Harvard University Press.
Darwin, F. (ed.) (1887). *The Life and Letters of Charles Darwin, Including an Autobiographical Chapter*. 3 vols. London: John Murray.
_____ and A. C. Seward (eds.) (1903). *More Letters of Charles Darwin*. 2 vols. London: John Murray.
Ghiselin, M. (2000). *The Triumph of the Darwinian Method*. Chicago: The University of Chicago Press.
Hodge, M. J. S. (1977). The Structure and Strategy of Darwin's "Long Argument." *British Journal for the History of Science* 10: 237-246.
_____ (1987). Natural Selection as a Causal, Empirical and Probabilistic Theory. In *The Probabilistic Revolution*. Vol. 2. L. Krüger (ed.). Cambridge, Mass.: The MIT Press, 1987, 233-270.
Kitcher, P. (1985). Darwin's Achievement. In Rescher, N. (ed.). *Reason and Rationality in Natural Science*. Lanham: University Press of America, 127-189.
Lakatos, I. (1970). History of Science and Its Rational Reconstructions. *Proceedings of the Biennial Meeting of the Philosophy of Science Association*, 91-136.
Lloyd, E. (1983). The Nature of Darwin's Support for the Theory of Natural Selection. *Philosophy of Science* 50: 112-129.
Mayr, E. (2004). *What Makes Biology Unique?* Cambridge: Cambridge University Press.

Okasha, S. (2000). Darwin. In *A Companion to the Philosophy of Science*. W. H. Newton-Smith (ed.). Oxford: Blackwell.

Popper, K. R. (1974). *Unended Quest*. London: Routledge.

___ (1978). Natural Selection and the Emergence of Mind. *Dialectica* 32.

Reed, E. S. (1981). Lawfulness of Natural Selection. *The American Naturalist* 118 (1): 61-71.

Ruse, M. (1975). Charles Darwin's Theory of Evolution: An Analysis. *Journal of the History of Biology* 8 (2): 219-241.

Sarton, G. (1937). Darwin's Conception of the Theory of Natural Selection. *Isis* 26 (2).

Sintonen, M. (1985). Darwin's Long and Short Arguments. *Philosophy of Science* 57 (4): 677-689.

Sober, E. (1984). *The Nature of Selection*. Cambridge, Mass.: The MIT Press.

Thagard, P. (1978). The Best Explanation: Criteria for Theory Choice. *The Journal of Philosophy* 75: 76-92.

Wilson, F. (1992). *Empiricism and Darwin's Science*. Dordrecht: Kluwer.

8
Revisiting Popper's Evolutionary Theory of Knowledge

Vikram Singh Sirola

Popper's innovative account of the origin and evolution of knowledge corresponds with the origin and evolution of life. He claims that his new theory of knowledge overturns everything that was said on the topic by his predecessors. This chapter attempts to develop Popper's evolutionary theory of knowledge by outlining how his attempt to take historicity seriously leads to a transformation of thought. And yet I shall argue that history of science does not bear out Popper's novel account of knowledge. His account is not false but incomplete and, in its present form, of a limited scope. I would go with his astute claims that "only with life do problems and values enter the world"; "two important values, critical approach and objective truth entered our world only with the human language" and; "theories are inventions of our own making." But I shall show that these claims are severely limited by his fundamental contention that physical codification of environmental conditions is knowledge. I shall try to show that the knowledge we gain through this codification is severely limited. It fails to account for the explorations made in modern science of the realms of reality not accessible to our everyday experience. The chapter attempts to develop an understanding of life as a cognitive process. An insight into the character of knowledge and the genesis of epistemic subject demands an account which goes beyond the information storing systems—cognitive structures and the environment.

Popper found the term "evolutionary epistemology" a bit pretentious. In his lecture delivered at the London School of Economics in June 1989, he said he preferred the expression "evolutionary theory of knowledge" (Popper 1999, 57). He does not give a primacy to the question, What is knowledge? Instead, he prefers to study our acquisition of knowledge, its growth and evolution in parallel with the evolution of life. His objective is to place the work in theory of knowledge in the wider context of biological evolution. Such exercise, he believes, might offer an exciting new description and explanation of our cognitive processes. In contrast to Popperian thesis the main philosophical use of evolutionary theory in recent years has been with respect to the character and development of human cognitive capacities. Darwin fills that long standing gap for him.

Popper's idea of evolutionary theory of knowledge, primarily developed in *The Logic of Scientific Discovery,* deals with the problems of methodology and scientific progress. He argues against the empiricist-positivist tradition with its emphasis on induction. He holds that scientific theories are not arrived at by means of induction. Collecting observations cannot yield knowledge. Positivists never closed the gap between observation and knowledge. Popper hopes to close this gap by drawing on biology. He notes that we always start from some hypotheses or theoretical preconceptions. In other words, observation is always permeated with some theory, however rudimentary. Theories are our inventions. Our imagination may draw on any conceivable source for the construction of theories, including myth and metaphysics. Thus, Poppers treats knowledge as non-representational and groundless.

Now, it may be asked, how can we be sure that our knowledge corresponds to the world, instead of being a mere figment of our imagination? Popper argues that while we cannot indeed be sure that our knowledge is ever true, we can find out that it is false. As he puts it, our theories cannot be verified, but they can be falsified. We have to submit our hypotheses to severe tests. While no amount of tests confirms a hypothesis, it can fail in a single test. A failed hypothesis must be discarded, one which passes can be retained, but only as long as it keeps passing future tests. The significant point here is that we can have knowledge about the world even though this knowledge is not achieved by means of induction. Moreover, the truth of that knowledge is not justified by the method used to gain it. Hence, sensory observations, protocol statements and induction are inept for the purpose.

Popper's theory of the growth of knowledge through falsification is widely believed to be a counterpart of Darwin's theory of evolution, in particular the latter's concept of the survival of the fittest. Yet the analogy between the two theories goes only so far. In particular, Darwin's conception of evolution through variation and natural selection does not admit saltations; evolutionary developments are gradual and slow, admitting only slight variations. By contrast, scientific progress as defined by Popper, proceeding through conjectures and refutations, envisages a rapid growth. John Watkins puts it aptly: "For Darwin there could be no such thing as a 'hopeful monster'; but the history of science, seen through Popperian eyes, is full of 'hopeful monsters'" (Watkins 1995, 193).

Popper compares adaptation to truth. There is no direct feedback between the environment and the organism, so no organism is ever perfectly adapted. Likewise, no theory can ever claim to be absolutely true. Moreover, the environment does not induce mutations which would help the organism survive. New mutations appear randomly, they are imperfect solutions to the problems faced by the organism with no regard for a possible success in a given environment. Similarly, our hypotheses in a scientific enquiry are not induced by experience. Hypotheses are rather inventions of our minds which have to stand on their own in the hostile world of experimental testing.

Popper's idea that we acquire knowledge through the process of trial and error elimination has surely been one of the great ideas of the twentieth century. This theory turns out to be vital features of Popper's comprehensive philosophy, especially in the later part of his life. It became fundamental idea behind his most accomplished contributions in methodology and epistemology. He offers it as a model to explain the growth of individual knowledge in micro-organisms or humans and also of scientific knowledge. The theory of trial and error elimination was built on rejecting of what Popper labels as, the "bucket theory of mind." His proposal is that our knowledge of the world is partly drawn from our mind and constructed from the repertoire of knowledge dispositions we already possess. These dispositions are mostly innate or acquired modification of what is innate.

This method applies to animal learning as well as to the scientific research. The idea goes like this: individual organisms encounter problems as soon as they are disappointed in their expectations, which are innate. Faced with the problems, organisms try out a number of solutions which Popper calls "testing movements." These movements are subjected to different internal and external conditions followed by process of error elimination. Learning here means that the false trials are gradually discarded and the successful ones are retained. Based on these, the organism forms new *expectations* that the problem can be solved by the one trial that has not been eliminated.

This model suggested by Popper has its parallel in Darwin's theory of the evolution of species. A species can survive only if it adapts through mutation in its genetic structure. Mutation here corresponds to Popper's "attempted solutions." The fit between organism and environment is decided by the success of the adapted trials. Organisms learn from nature but without instructions. They don't pick up and retain information from the environment. It further gets validated by the central dogma in molecular biology that DNA can instruct the proteins of which organisms consist but not the other way round. Acquired characteristics are not hereditary and do not play any role in evolution.

Popper makes use of the biological notion of adaptive success to claim that scientific knowledge is an organizing instrument in the organism's struggle to maintain its existence, to invade and even to invent new environmental niches. His model of trial and error elimination continues to apply at this level too. Dominant theories are continuously exposed to new theoretical problems. The attempt is to look for an increased fit between scientific theories and facts of nature. Scientific knowledge progresses by conjectures and hypotheses which are always susceptible to elimination. Progress or acquisition of knowledge in science takes place by a negative process of elimination. It would not seem an exaggeration to say that organisms are embodied theories about the environment, and that theories held by conscious human beings are disembodied organisms.

Franz Wuketits differs from Popper and proposes a non-adaptationist approach to evolution. For him, what is important is "not how animals and humans have evolved through adaptation to a given environment, but rather how the interactions between organisms and their environment(s) have evolved" (Wuketits 1995, 359). Instead of a correspondence between cognition of the organism and the external world he argues for coherence between the two. Our perception of the outer world is not necessarily an objective image of it. Instead reality for us is our view of it which rests on coherence and success in life. This systems-theoretic approach removes the duality between the organism and its environment and proposes a view of organic evolution.

Popper's theory of trial-and-error elimination gives primacy to problems or expectations over observations. He reversed the role of sense perception as envisaged by the empiricist tradition. Knowledge of the external world does not originate by sense perception. Problems and expectations act as a searchlight which enables the organisms or the scientists to know what to observe. Even our sense organs may be seen as the outcome of a series of biological problem solving process from an evolutionary point of view. In this framework knowledge and experience are being constructed from the built-in repertoire of expectations and dispositions rather than being the passive outcome of the collection and association of sense perceptions. Popper declares the empiricist's theory of knowledge a myth. As an alternative he proposes a revolutionary theory of knowledge which, he says "overturns everything my predecessors have said up to now. We are active, we are constantly testing things out, constantly working with the method of trial and error." His epistemological notion of genetic *a priori* knowledge accentuates the role of inborn knowledge but, unlike other nativists, conceives of this knowledge as fallible, provisional, and continually subject to refutation on empirical basis.

In *Logic of Scientific Discovery*, the topic of a growth of knowledge is the central problem of epistemology (Popper 1959, 15). In *Conjectures and Refutations*, the solution to the methodological problem of demarcating science from non-science is at the centre. Popper finds it as "the key to most of the fundamental problems of the philosophy of science" (Popper 1963, 42). In later works, his evolutionary hypothesis gets more concerned with "understanding human knowledge as continuous with animal knowledge; and to understand also its discontinuity—if any." He examines the whole realm of cognitive structures found in the animal kingdom, and compares the "fit" between the organic system and its environment. His conception of knowledge is based on anthropomorphism and theory of homology. Animals behave dogmatically in the sense that without waiting for premises they jump to conclusions to which they stick even when the evidences point towards the contrary. Thus, Popper significantly generalizes his approach: our experience is theory-impregnated and structure-impregnated. He recommends a critical attitude to learning by trial and error, which consists of giving up the dogmatic holdings, ones faced with disappointed expectations and then attempting newer solutions. This readiness to test and change expectations, although present in the learning activity of most organisms, is perfected in science. Accordingly, Popper equates critical attitude with scientific attitude. He explains a difference between science and a pre-scientific problem solving by means of a comparison between Einstein and an amoeba: "Although both make use of the method of trial and error elimination, the amoeba dislikes to err while Einstein is intrigued by it: he consciously searches for his errors in the hope of learning by their discovery and elimination." Though Einstein is only a step ahead of amoeba, it is a significant step enabled by the existence of language. Only language makes it possible to evaluate our theories critically as we can look at them as external objects belonging to a world outside of ourselves and shared with others.

We cannot overlook that an appeal to Darwin's theory of evolution generates a kind of epistemological pessimism. Nature, in its selective process of evolution, has fitted us with certain forms of learning and understanding that help increase an individual organism's survival and its chance of reproduction. It can be argued that what is vital in survival and reproduction does not include deep and profound human intellectual accomplishments. It is not expected of us to fathom the origins of life, mysteries of universe, or the nature of consciousness. Yet philosophers and scientists claim that humans can understand some of these things. Hence, the Darwinian proposal seems to have a limited application.

In naturalized epistemology, Darwin's theory is used as the footing for the claim that science has severe limitations. These limits are drawn by the evolutionarily derived frailty of human understanding. It is noteworthy to see how Popper, one of the proponents of naturalized epistemology, would respond to this epistemological pessimism.

By drawing a parallel between knowledge and the Darwinian evolution, Popper gave a distinctive turn to evolutionary epistemology to explain the growth of knowledge. Nicholas Rescher calls it the "Thesis Darwinism" (Rescher 1990, chap. 2). Popper's Thesis Darwinism strikes a parallel between his method of conjecture and refutation, on the one hand, and the Darwinian trial and error method, on the other. Popper writes:

> the growth of our knowledge is the result of a process closely resembling what Darwin called "natural selection"; the natural selection of hypotheses: our

> knowledge consist, at every moment, of those hypotheses which have shown their (competitive) fitness by surviving so far in their struggle for existence; a competitive struggle which eliminates those hypotheses which are unfit. This interpretation may be applied to animal knowledge, pre-scientific knowledge, and to scientific knowledge. What is peculiar to scientific knowledge is this: that the struggle for existence is made harder by the conscious and systematic criticism of our theories. Thus, while animal knowledge and pre-scientific knowledge grow mainly through the elimination of those holding the unfit hypotheses, scientific criticism often makes our theories perish in our stead, eliminating our mistaken beliefs before such beliefs lead to our elimination.
>
> This statement of the situation is meant to describe how knowledge really grows. It is not meant metaphorically, though of course it makes use of metaphors. The theory of knowledge which I wish to propose is largely Darwinian Theory of growth of knowledge. From the amoeba to Einstein, the growth of knowledge is always the same. (Popper 1973, 261)

There are two important contentions which could be drawn from this proposal. One, the theory-elimination method is seen by Popper as a progressive enterprise getting us closer to the truth. But Darwinian idea of evolution, on which it is based, is non-progressive and doesn't make such claim. Secondly, using this to understand the scientific enterprise equips us to see "how it is" not "how it ought to be." It merely puts forward the way or at best provides us with the explanation of how science is being practiced. Scientific claims would always be provisional, inconclusive, and fallible. As Popper says, we cannot produce an adequate solution to any problem—practical or theoretical. The growth of knowledge and a progress towards the truth involves the succession of inadequate solutions and their fierce criticisms. At best, this theory explains the very criteria of predictive success used in science to judge the authenticity of theories.

Nevertheless, it is equally significant to note its limited application. It has been proven now that predictive success in a particular field of inquiry may be achieved in different ways delineated by mutually inconsistent theories. Consider an example—a choice between Fresnel's wave theory of light and Maxwell's electromagnetic theory. These theories are based on a mutually inconsistent understanding of the nature of light. But they stand equal with respect to predictive success as they share a common mathematical structure. John Worrall explains:

> Roughly speaking it seems right to say that Fresnel completely misidentified the nature of light, but nonetheless it is no miracle that his theory enjoyed the empirical predictive success that it did; it is no miracle because Fresnel's theory, as science later say it, attributed to light the right structure. (Worrall 1989, 117)

Our view of the world has changed considerably over the last few decades. For very long we had believed that only human culture had a history and that it was not part of nature; nature, in its turn, was seen as continuous, unchanging and ahistorical. This perception has been corrected by Darwinism. Yet is it not a drawback of Darwinism that it conceives of nature and science as belonging to a plane devoid not only of human subjectivity, but also of cultural, aesthetic and religious experiences, and their historical dimension? Popper claims to be aware of this fact but fails to overcome it. The evolutionary view cannot take into consideration what is vital to the human social life, to that which is higher and directs life towards its perfection. And, importantly, none of what is higher can be claimed to be indispensable to men's survival—directly or indirectly. For example, the best astrophysical theory may not have any connection with a success in survival.

Popper argues in *Realism and the Aim of Science* that the values and beliefs of a social group may be seen as having a physiological basis. It is a kind of bondage "imposed by us upon the world, in the sense that they are bound to become 'objective' laws of the things which we perceive." But these barriers of culture and biology, he claims, we can overcome by

> learning to criticize ourselves and so to transcend ourselves. [...] Rational discussion and critical thinking are not like the more primitive systems of interpreting the world; they are not a framework to which we are bound and tied. On the contrary, they are the means of breaking out of the prisons—of liberating ourselves. (Popper 1983, 154-155)

However, our beliefs and values would be very limited notions, if defined in terms of their adaptive use. Then even our greatest intellectual achievements might turn out to be, as suggested by Michael Ruse, "illusions fostered upon us for reproductive purposes." Popper overlooks that we are epistemologically "over-reachers" (Smithurst 1995, 213). Human intellectual capacity constitutes a case against the evolutionary view. Thomas Nagel also argues that "if, *per impossibile*, we came to believe that our capacity for objective theory were the product of natural selection, that would warrant serious scepticism about its results beyond a very limited and familiar range" (Nagel 1986, 79).

Scientific problems do not present themselves as the problem features of the world. Rather, they are our projective concerns. Scientific information is an "ideational construct"; the sameness of the object does not guarantee the sameness of thoughts about the object. This view could be vindicated by the discovery of creatures, perhaps aliens, with a different bodily and mental makeup, that could interact with us in this world. These alien creatures would deal with the same world as ours, but might differ in their mode of conceptualisation. If developed and rational enough, such life forms might cultivate their own kind of science. It would then be appropriate to conceive of natural science as providing a picture of "reality as it presents itself to us," rather than as picturing "reality as such."

When we think of candidates for the status of a rational agent pursuing a scientific knowledge, we seem to assume that such an agent must be either a human being or someone very much like it. As Wittgenstein writes, "Only of a living human being and what resembles (behaves like) a living human being can one say: it has sensations; it sees; is blind; is deaf; is conscious or unconscious" (Wittgenstein 1953, §§ 360 and 281). It is true that our "ways of speaking" about the world cannot receive any justification for some sort of non-linguistic knowledge. It seems radical to base the harmony between thought and reality entirely on the rules of our grammar. But Wittgenstein does make a very significant point here. The most important epistemic values—self-criticism and the concern for truth—did enter the (human) world only with the emergence of human language. It is in this sphere that our problems, values, and activity all evolve and grow together as an organic whole.

Finally, I shall make two concluding points. First, are we now in a position to answer the question, What is the purpose of human life? Does this question have any meaning within the evolutionary framework, or should it be dismissed? One may offer a cynical and frivolous answer to this question by suggesting that, within the Darwinian paradigm, all life has the sole purpose of self-preservation at any cost. I suppose Popper would reject such an answer. Yet it is difficult to justify a hierarchy of values within the Darwinian paradigm.

Second, Popper's core thesis is that science advances by conjecture and refutation. Conjecture is a matter of the unrestrained use of imagination in a non-deterministic world. Popper argues that determinism should be dismissed, as it is incompatible with the very possibility of creativity. True novelty is impossible in a deterministic system. Popper makes an important point here: "Intelligence is arguably a faculty, arguably a unitary capacity, to some degree inherited, and with a metric assignable to it." But this is not a plausible way to conceive of creativity. According to Popper, conceptual and explanatory innovation emerges from a confrontation with problems, unforeseen and unforeseeable problems. Creative imagination is not such a latent potentiality. It originates in the interaction of a creature with its environment. It is not an unfolding of some dormant thoughts. It is true that Popper claims that "the theory that the creation of works of art or music can, in the last instance, be explained in terms of chemistry or physics seems to me absurd" (Popper 1982, 127–128). However, this claim becomes a mere rhetoric if creativity is to be seen holding with *a priori* grounds. Within his framework, it can be shown that human thoughts must inevitably take certain predictable forms, as for him knowledge is incorporated in our biochemical constitution. Even the inventions are seen as incorporated in the structure of the organism; in a new inheritable knowledge, and therefore in new *a priori* knowledge (cf. Popper 1999, 70).

References

Nagel, T. (1986). *The View Form Nowhere*. Oxford: Oxford University Press.
O'Hear, A. (1995). *Karl Popper: Philosophy and Problems*. Cambridge: Cambridge University Press.
Popper, K. (1959). *The Logic of Scientific Discovery*. London: Hutchinson and Co.
_____ (1963). *Conjectures and Refutations*. London: Routledge.
_____ (1973). *Objective Knowledge: An Evolutionary Approach*. London: Oxford University Press.
_____ (1982). *The Open Universe*. London: Hutchinson and Co.
_____ (1983). *Realism and the Aim of Science*. London: Hutchinson and Co.
_____ (1999). *All Life is Problem Solving*, London: Routledge.
Rescher, N. (1990). *A Useful Inheritance: Evolutionary Aspects of the Theory of Knowledge*. Savage, Maryland: Rowman & Litlefield.
Smithurst, M. (1995). Popper and the Scepticisms of Evolutionary Epistemology, or, What Were Human Beings Made For? In O'Hear (1995), 207-224.
Watkins, J. (1995). Popper and Darwinism. In O'Hear (1995), 191-206.
Wittgenstein, L. (1953). *Philosophical Investigations*. Oxford: Blackwell.
Worrall, J. (1989). Structural Realism: The Best of Both Worlds? *Dialectica* 43: 99-124.
Wuketits, F. M. (1995). A Comment on Some Recent Arguments in Evolutionary Epistemology—and Some Counterarguments. *Biology and Philosophy* 10: 357-363.

9
Criticism and Dogmatism in Popper's Evolutionary Epistemology

Zuzana Parusniková

Karl Popper was a philosopher with a broad range of interests. In his younger years in between the wars he was influenced by logical positivism, a dominant philosophical movement in Vienna in that time, and participated in the some of discussions taking place at the outskirts of the Vienna Circle, especially those focused on problems of natural science, rationality, logic and methodology. The political climate before the Second World War and especially the tragic consequences of the rise of Nazism made Popper look at philosophical issues in the social sciences, and think about various ideologies encouraging the rise of authoritarian, totalitarian tendencies in societies. Against these, he put forward the concept of the open society in which he appealed to strengthening the democratic liberal pillars of our Western culture. After the war, at the beginning of his career at LSE, Popper first worked on both the political and scientific themes but gradually became ever more preoccupied by epistemology, drawing on his early interest in the growth of knowledge. And this preoccupation inspired him to consider a parallel between his theory of the growth of knowledge and the Darwinian theory of natural selection. Popper was one of the founders of nowadays very influential discipline known as evolutionary epistemology thus opening the way to the explanation of knowledge as an integral part of the evolution of life on Earth.

1 Evolutionary Epistemology

The term "evolutionary epistemology" is not, as is well known, Popper's own creation but was introduced by Donald Campbell, a psychologist, anthropologist and ethologist who viewed the development of knowledge in a similar perspective; in this context, he was mainly interested in how our sensory and perceptive capacities can develop into creative thinking capacities. Campbell valued highly Popper's contribution to the emerging new discipline—in his own words, "it is primarily through the works of Karl Popper that a natural selection epistemology is available today" (Campbell 1974, 413). He correctly pointed out that Popper's epistemology is based on his rejection of the model of learning by passive induction—instead, learning follows the model of trial and error elimination. Bold conjectures aspire at truth, and reality itself (as it reacts to empirical tests designed to falsify the conjectures) decides which of them will be eliminated and which retained. It is the discovery of errors that opens up new areas of problems and thus drives knowledge forward.

Campbell emphasized the fact that Popper's evolutionary model is universal:

> the problem of knowledge is so defined that the knowing of other animals than man is included. The variation and selective retention process of evolutionary adaptation is generalized to cover a nested hierarchy of vicarious knowledge processes, including vision, thought, imitation, linguistic instruction, and science. (Ibid., 450–451)

What unifies human and animal learning is that they are *various forms* of essentially *one process* of problem-solving. Problems to be solved may be threats to survival, urgent practical challenges of the surrounding world, or theoretical challenges posed by errors in our proposed explanations of the world. "From the amoeba to Einstein," as Popper says, "the growth of knowledge is always the same: we try to solve our problems, and to obtain, by a process of elimination, something approaching adequacy in our tentative solutions" (Popper 1979, 261). Popper does not hesitate to call the adaptation processes taking place in primitive forms of life a form of knowledge and thus "the origin of the evolution of knowledge may be said to coincide with the origin of the evolution of life and to be closely linked with the origin and evolution of our planet earth" (Popper 2003, 64).

And although Popper began to draw explicit parallels between natural selection and the development of knowledge later on, in the 1960s and 1970s, the basic pattern can be found already in his *Logic of Scientific Discovery;*

> Its aim (of falsification) is not to save lives of untenable systems but, on the contrary, to select the one which is by comparison the fittest, by exposing them all to the fiercest struggle for survival. (Popper 1959, 42)

The struggle for survival is a metaphor expressing the process in which of our theories aspire to hold on to the status of being true—and with this aspiration they enter an arena of ruthless competition determined by the rule that all unsatisfactory theories are eliminated.

Already during his Vienna years, Popper inclined to emphasize the dynamic, procedural aspects of science; by contrast, the focus of the Vienna Circle was predominantly on strict and rigid definitions of the demarcation criterion between science and non-science, and of the logical relations holding between generalizations and empirical data. Popper, however, held the view that "[t]he central problem of epistemology has always been and still is the problem of the growth of knowledge" (Popper 1959, 15). It is also for this reason that he embraced the principle of falsifiability; that is, openness to falsification. Falsification is not just the only rational (logically kosher) procedure of testing but it is also highly stimulating for discovering new problems and new solutions. Popper therefore criticized the probability logic used by the philosophers of Vienna Circle (and especially by Reichenbach), not just for logical reasons but also for epistemological reasons—the ideal of high probability of our theories leads to stagnation. The aim to achieve the highest possible probability of our statements inhibits our willingness to risk, and stifles the boldness and creativity of our thinking. Instead, unchallenging, banal, boring statements with low empirical content that can pass in tests are favoured. In my view, these themes concerning the dynamic aspects of cognition shaped Popper's philosophy ever after and finally inspired him to explore the extent to which Darwinism can provide an explanation of the development of science and of all human knowledge.

A quick glance at the main principles of Popper's Darwinian model should start with the famous evolutionary scheme (Popper 1979, 287):

$$TT_a \rightarrow EE_a \rightarrow P_{2a}$$
$$P_1 \rightarrow TT_b \rightarrow EE_b \rightarrow P_{2b}$$
$$TT_n \rightarrow EE_n \rightarrow P_{2n}$$

where *P* stands for a problem, *TT* for tentative solutions, and *EE* for error elimination. Cognition, then, starts with an objective problem, and proceeds from one conjecture to another by eliminating errors, enabling us to detect new problems and discover new areas of knowledge or, rather, new areas of ignorance.

As I said before, Popper applied this scheme to the whole area of life. Solving problems and detecting mistakes in our tentative solutions is not just a theoretical business—all organisms constantly strive to solve survival problems and their tentative solutions may be various adaptation strategies made instinctively in order to remove the threats and to adjust their behaviour or, in long term, their genetic set-up to deal better with the challenges of reality. Success in this endeavour is not, however, a once-for-all victory. The environment changes—and the adaptation strategies, too, contribute to these changes—and thus new unexpected problems arise. Active alterations of the environment are characteristic especially for humans—for them, the ability to create World 3, the world of objective knowledge, enables the formation of "care-free" critical activity shorn of the fear of personal death in case of a mistake, and thus boosts the growth of knowledge. And since knowledge is his biggest evolutionary advantage, man is, thanks to the capacity of reason (identified by Popper with criticism), a much more active player in the evolutionary game, having at his disposal more effective ways of solving problems and actively shaping his life-environment.

Despite these privileges, the basic evolutionary pattern is in Popper's view the same for all living creatures. As he says:

> from a biological or evolutionary point of view, science, or progress in science, may be regarded as a means used by the human species to adapt itself to the environment: to invade new environmental niches and even to invent new environmental niches [...] [w]e can distinguish between three levels of adaptation: genetic adaptation; adaptive behavioural learning; and scientific discovery [...] *[o]n all three levels, the mechanism of adaptation is fundamentally the same.* (Popper 1973, 78-79; emphasis in the original)

Yet, in human cognitive activity the phase of error-elimination is upgraded to an unprecedented status. In case of animals, extravagant errors result in physical death—the elimination of errors occurs in the World 1, the world of physical entities. According to Popper, humans have another option—to let theories die in their stead. The evolution of knowledge characterizing the evolution of the human species takes place in the world of objective ideas (World 3) and we can delegate the survival struggle to our ideas. True, this option is not always favoured and we still keep killing each other in pursuit of ideological goals, but the availability of the choice itself is unique to humans. And the right choice implies a positive attitude to error—since we do not fear for life we can actively pursue criticism and embrace its stimulating effects. To use Popper's example of Einstein again, "the difference between the amoeba and Einstein is that, although both make use of the method of trial and error-elimination, the amoeba dislikes erring while Einstein is intrigued by it: he consciously searches for his errors in the hope of learning by their discovery and elimination" (Popper 1979, 70). Erring (and only erring, in Popper's account) leads to new problems, stirs our curiosity, points out new directions of inquiry, and gives the promise of new discoveries. Popper vehemently emphasized this aspect of cognitive evolution and provocatively argued that, to the contrary, justification in any form contributes nothing whatsoever to the growth of knowledge. In knowledge, risk is worth it—no blood loss and huge intellectual compensation.

In a nutshell, despite the unifying underlying pattern "trial-error," the biggest difference between the biological and cognitive evolution—"lies in the constructive attitude towards error" (Popper 1963, 52). But this difference places humans on a qualitatively higher level of evolution than the rest of the natural world; it shows that only humans have at their disposal the capacity of reason, linked to the argumentative function of language and abstract thinking, and thus can develop the imaginative, reflexive and critical forms of intelligence. The transfer of natural selection from the physical to the cognitive area therefore significantly speeds up the evolutionary processes in our human world.

In his evolutionary epistemology, Popper tried to bring together several principles that are constitutive of his philosophy. The fact that evolution proceeds from one problem to another reflects Popper's rejection of induction as a mode of learning; this is explained in detail in his solution of the Hume's problem of induction. Induction must be rejected not only in logic, as Hume rejected it, but in the theory of knowledge acquisition. In this way too, Popper distanced himself from the psychologism of the Vienna Circle.

Further, the driving force of this advancement from one problem to another is criticism. In his logical analyses, Popper claimed that empirical falsification of statements is the only admissible (e.g., rational) procedure in testing. And testing is the only domain of rationality—reason cannot be seen as a controller of the way in which we construct theories, invent hypotheses. It is the controller of the selection process in which false hypotheses are eliminated. But as I said above, falsification does not just change our traditional view of the logic of discovery but highlights the dynamics of the process of learning. In a broader context, falsification can be replaced by the selection concerning, beyond theories, the whole of Nature. The imperative of falsification is thus perfectly in tune with the universal mechanism of evolution.

And finally, Popper's radical anti-justificationist stance denies the traditional epistemological value of certainty; Popper breaks away from the deeply rooted Cartesian tradition in which the capacity of reason to justify and hence establish the certainty/truth of our knowledge is the ultimate task; for him justification is both irrational and unproductive. He rejects the whole concept of striving for certainty, or probability, or reliability in the assessment of our guesses, and revives the neglected Socratic idea of humility, by which reason reminds us of our limitations, not of our achievements. Experience can be used only to expel false theories; and as for unfalsified theories "it provides no positive support for them, provides no inductive lift, leaves them floating in the ocean of uncertainty" (Watkins, 1984, 353-354). Unfalsified theories can retain their claim at the truth but their truth can never be proved. Popper asks us to exploit the positive potential of uncertainty and appreciate the fact that within the domain of uncertainty life is more open, adventurous, and free. This concept, I think, fits very well the Darwinian perspective on evolution as an open-ended process in which hitherto existing success does not contain any guarantees for the future, and the actual evolution of a species may take surprising turns.

Popper gives in his conception a comprehensive account on various aspects of both natural and cognitive evolution based on scientific research and cooperation with scientists—in this brief summary I have outlined only a few elementary features of it. His contribution to evolutionary epistemology has been widely acknowledged (see, e.g., Campbell 1974, Radnitzky and Bartley 1987, Munz 1985, Hahlweg and Hooker 1989, Wuketits 1990).

Despite this recognition, many objections have been raised against Popper's position. They can be roughly divided into two main groups: *firstly*, the question has been asked whether his evolutionary model is really fully consistent with Darwinism, and *secondly*, many philosophers distance themselves from viewing the development of knowledge purely in terms of selection as opposed to instruction.

The most notorious argument belonging to the first group of objections touches on the regulative ideal of truth in Popper's philosophy, suggesting the possibility of progress in approaching the truth. For Popper, truth is the ultimate goal of cognition; science, due to its most conscious and active application of criticism, offers the best way to achieving it. He says: "science is one of the very few human activities—perhaps the only one—in which errors are systematically criticized and fairly often, in time, corrected. This is why [...] we can speak clearly and sensibly about making progress there" (Popper 1985, 216). Our current theories are thus closer to truth than their predecessors.

Although Popper stresses that the future course of the development of knowledge is open, and that we can speak about progress only as far as the present state of knowledge is concerned, there is, I think, a strong Enlightenment streak in his philosophy. He assumes that knowledge has a liberating mission in realizing progress (getting closer to the truth) both in the sense of mastering Nature and achieving a more civilized forms of society, and that during history our capacity for critical thinking and therefore also our capacity to realize progress intensifies. In any case, defining truth as the goal of the evolution of knowledge, and interpreting the actual course of the evolutionary process up to now as partially fulfilling this goal, opens the door to the orthogenic perspective on the growth of knowledge—a perspective that is incompatible with Darwinism in which biological evolution consists of random mutations resulting in accidental variations of life forms with no teleology involved (as Popper himself frequently claims). This would indicate an inconsistency in Popper's views undermining the Darwinian nature of his epistemology.

But the situation is not so clear-cut. Teleology is a broad concept and can be interpreted in a massively metaphysical sense implying an intelligent design in Nature and assuming that there is a final goal unfolding itself through the development of life (with knowledge being an integral part of it). But it can be also interpreted more "modestly" as a tendency to follow a natural law. Natural scientists, especially evolutionary biologists, commonly work with this concept; for instance Grehan argues that in biology, teleology of this kind is accepted and expresses the fact that "change in form is merely a consequence of the initial state, followed by evolution according to laws and potential" (Grehan 1984, 16). With advancing scientific research, interesting implications of this low-level teleology have been discussed in genetics, looking at the nature and role of mutation. The mechanism of "molecular drive" in cells (Dobzhansky 1970, Berry 1982) gives recognition to a possible directional quality in evolution.

But even Darwin himself may be considered to have employed a certain teleological concept and to have inclined towards higher-caliber teleology. In his law of growth he attributed evolution to directional factors independent of the environment and clearly indicated that natural selection came to work on some inbuilt tendency, some drive toward perfection—these "residues" of the theological vocabulary (the influence of Paley's *Natural Theology* on Darwin) are totally rejected by the Neo-Darwinian orthodoxy. But even without this particular aspect of Darwin's theory one could ask whether a certain metaphysical element does not lie merely in the principal assumption of the universal will to life inherent in living creatures. Without it, there would be no struggle for survival as the basis of evolution. The issue of teleology in Darwinism and in the Neo-Darwinian evolutionary theory thus appears much more complicated and should be examined within the whole variety of different aspects.

Secondly, Popper proposes that the growth of knowledge proceeds via ruthless critical selection of hypotheses—on the biological level this corresponds to the principle that evolution proceeds via natural selection in which only the fittest survive. In order to have the flexibility to adapt slowly and gradually organisms within populations exhibit individual variations—these arise by mutation, a change in some part of the genetic code for a trait. In the evolution of knowledge, however, there has been a tremendous "rockety" boom over the last few centuries, having no parallel in any changes of our genetic make-up, not transmitted by inheritance and driven not by blind guesses but by conscious and systematic efforts to fill certain well-defined holes in current knowledge. The evolution of knowledge, therefore—some philosophers argue—proceeds by instruction rather than selection.

Popper attempts to accommodate some of these criticisms mainly by distinguishing between instruction from within (learning from existing knowledge) and from without (from the outside environment) and allowing only the former. He concludes:

> [...] the critical or Darwinian approach allows only *instruction from within*—from within the structure itself [...] *there is no such thing as instruction from without the structure*, or the passive reception of a flow of information that impresses itself on our sense organs. (Popper 1973, 84)

Popper's solution, however, is debatable. Why does Popper consider existing knowledge as being the source of instruction from within? After all, he brought into epistemology the original concept of the world of objective knowledge having an ontologically independent status, being a world of its own, living its own life. In epistemology, this structure *is* the outside environment representing, as Popper claims, a collection of problems that challenge us and motivate the cognitive subject to pursue certain lines of inquiry. Therefore, if we learn from this structure by instruction, it is learning from without and if Popper claims the opposite he contradicts his previous definitions in which the only epistemological benefit of World 3 consists in provoking us to resolve problems by blind guessing.

The argument that there is an element of instruction from without also implies that Popper underestimates the extent to which theorizing is guided by rational considerations (as opposed to blind guesses) (O'Hear 1999, 53-57). Similarly, using the example of Fresnel's wave theory of light, Worrall claims that in science, the heuristic guidance provided by background knowledge is much stronger than Popper assumes. Scientists are thus instructed from the epistemic environment (from without) and new hypotheses arise largely due to systematic, deliberate, "controlled" cognitive activity (Worrall 1995, 90-102).

This brief excursion into Popper's evolutionary epistemology and its problematic aspects certainly does not do justice to the complexity of his ideas on this subject. My aim was to highlight at least the key ideas behind his evolutionary model and to show the main directions in which critical debates have been carried out. This summary enables me to place my own critical comments in the appropriate context. In my argument I will look at the role that Popper attributes to criticism and to dogmatism in his epistemology, and show that the set-up he proposes causes a tension—if not an inconsistency—with Darwinism.

2 Criticism, Dogmatism and Evolution[1]

The rejection of induction as a mode of learning is the pillar of Popper's selectionism in epistemology. In the inductivist epistemology cognition begins by observation when information is received through our senses and is then accumulated and "digested" through various cognitive procedures (the bucket theory). Popper replaces this inductive pattern by a deductive frame in which learning begins with problems that determine the perspective and the range of our observations (the searchlight theory). All observation is theory impregnated and all proposed solutions to the initial problems are tested with aim to refutation. On the biological level one can speak instead of initial cognitive problems of inborn expectations about the world that, once confronted with a reality that does not satisfy them, brings forward practical problems; these expectations trigger off the process of forming experience and thus lie at the basis of the cognitive progression.

This solution, however, is a source of problems that Popper did not foresee. Namely, our inborn expectations have a dogmatic nature since they contain a

---

[1] Parts of this section are drawn from Parusniková (2004).

strong need for confirmation, encouraging us to justify rather than falsify our beliefs. In other words, we are dogmatists by nature and this instinctive outfit forms our cognitive attitudes. But according to Popper, the growth of knowledge is driven by active criticism (error-elimination)—this is the key characteristic of both his methodology and his epistemology. My main argument takes off from here: *in Popper's account, our spontaneous (dogmatic) cognitive inclinations contradict the objective (critical) pattern of the growth of knowledge.* Popper simultaneously holds both these views without fully realizing the full implications. *Our dogmatic cognitive inclinations are produced in the evolution and yet they sabotage the evolution of knowledge*, although the growth of knowledge plays a decisive role in the success of the evolution of mankind. Popper places these two mortal enemies (dogmatism and criticism) side by side in the evolutionary game—while their hostility is well accounted for in his methodology (totally denying dogmatism the right to existence) in epistemology it causes a friction. I shall try to show how this friction arises from Popper's own solution to the psychological problem of induction and how it undermines the consistency of his evolutionary epistemology.

Let's start with a brief reminder of his elimination of psychologism. This brings us to Hume. Popper praises Hume for his solution of the logical problem of induction but criticizes him for not having gone far enough, for not having considered that induction plays no role in belief formation. Thus, Popper concludes, Hume buried "the logical gems in the psychological mud" (Popper 1979, 89). Rejecting Hume's associationist psychology is the crucial step that Popper makes in rejecting the existence of inductive learning. According to Hume, people believe that instances of which they have no experience will conform to those of which they have experience. This belief is a result of repeated experience, where we observe that certain things are always connected (either in conjunction or in succession). The association mechanisms of our mind make us believe that these connections will continue to hold in the future. Popper claims that "induction—the formation of a belief by repetition—is a myth" (1979, 23).

In this critique of Hume, Popper establishes his epistemological position, according to which knowledge does not start with observation but with conjectures that act as filters of observation. Conjectures embody certain expectations that we have about the world and navigate, like a searchlight, the focus and selection of observation—learning consists in bold proposals and attempts to falsify them empirically. These expectations, Popper stresses, are not a result of inductive generalization, but are guesses imposed upon the world, often without any prior experience—"expectations may arise without, or before, any repetitions" (ibid., 24). What then, one may ask, gives the impulse for the expectations—at least for the very first initialising expectation that sets the cognitive process in motion—to arise? Couldn't that be some preceding passive perception? And isn't there a danger of an infinite regress lurking, or isn't this the kind of an insoluble question of what was first, hen or an egg?

Popper rejects this possibility and claims that the hen and egg question is soluble in his epistemology: "[g]oing back to more and more primitive theories and myths we shall in the end find unconscious, *inborn* expectations" (1985, 47). The original frame of reference determining observation, the very first stimulus of the learning process consists in *apriori* biological dispositions that are part of our genetic hardware. By this step, Popper reduces the psychological problem of induction to a biological one: "[...] psychology should be regarded as a biological discipline" (1972, 24), and subsequently solves it by genetic *apriorism*.

If we look closer at how Popper defines such inborn expectations, we may be disappointed. Popper's description is poorly argued and lacks scientific rigour. In what resembles a personal narrative, Popper names only one inborn expectation: "it was first in animals and children, but later also in adults, that I observed the immensely powerful *need for regularity*..." (ibid., 23). This expectation of regularities, however, brings dogmatism in the front line—Popper argues that *the belief in regularities is dogmatic* (1985, 49):

> Our propensity to look out for regularities, and to impose laws upon nature, leads to the psychological phenomenon of *dogmatic thinking* or, more generally, dogmatic behaviour: we expect regularities everywhere and attempt to find them where there are none; [...] and we stick to our expectations even when they are inadequate and we ought to accept defeat.

He disagrees with Hume's explanation of how this belief in regularities *arises* and thus solves the psychological problem of induction. But another big problem pops out: we are genetically conditioned to expect (even impose) regularities, e.g., that the sun will rise tomorrow—even if, say, we've seen it rise only once (or never) before. In a similar way, we make many other conjectures and expect them to hold. Moreover, we desire to find them confirmed and believe that positive empirical evidence is an attribute of confirmation justifying their acceptance. This attitude is dogmatic and is genetically imprinted on human beings.

Not only, then, is there a dogmatic streak in our nature, but it is a primal instinctive force. As Popper says, "this dogmatic attitude, which makes us stick to our first impressions, is indicative of a strong belief; while a critical attitude [...] is indicative of a weaker belief" (ibid.). And since the world around us is so diverse that it is easy to find a confirmation for almost any statement, if that is what we are really after, dogmatism has perfect conditions to flourish.

The solution of Hume's problem is achieved at a high price—its by-product is establishing *dogmatism as an objective biological determinant of our cognitive attitudes*. Popper classifies as *apriori* the dogmatic inclination to confirm our expectations. *We are natural-born justificationists and dogmatists*, just as amoeba and other species. I suspect that Popper did not acknowledge the full implications of this philosophical position. In order to show that knowledge does not begin with observation, he also established an antagonism between our dogmatic (justificationist) nature and critical (falsificationist) reason. While this is fine in methodology, it raises questions in evolutionary epistemology. Our instincts command: "confirm conjectures" while reason sets the norm: "refute conjectures." This antagonism implies that in Popper's epistemology, *our cognitive instincts sabotage the objective (selectionist) pattern of the growth of knowledge that proceeds via conjectures and actively sought refutations.* They undermine the effective development of our rational abilities, defined by Popper as directly contradicting and totally incompatible with any dogmatic or justificationist moves.

Why did humans develop the wrong cognitive instincts—instincts that are counterproductive for their own survival? The key to the successful survival of the human species is the growth of knowledge stimulated by criticism. Yet, our instincts—themselves a product of evolution—obstruct the active, eager search for refutations, and thus suppress the growth of knowledge. Instincts do not welcome conflicts between our expectations about reality and reality itself. We instinctively yearn for our expectations or hypotheses to be confirmed, and we believe that empirical support does the trick. Some people have stronger dogmatic tendencies than others—are keener justificationists than others—but they all share the same basic instinct. But this instinct stifles the dynamism of the growth of knowledge that occurs only through conflict and disagreement, unearthed by constant critical probing that undermines our natural (but irrational) attachment to a safe, predictable, and stable world.

As a result, the biological evolution and the evolution of knowledge work against each other. The logic of Popper's arguments itself suggests that we (and amoeba) are forced into error-elimination by external pressures, in order to survive. Humans can carry out this procedure on a higher, theoretical level, but *not even they do it willingly or eagerly*. There may be a few exceptions, like some scientists who are completely obsessed by undogmatic, open-minded problem searching—and not surprisingly they are usually viewed as nuts, cut off from the Humean common life. *Criticism is painful and contradicts our nature.* Searching for and admitting mistakes is done grudgingly and against strong dogmatic resistance (especially when our own mistakes are in question). This is a conclusion that Popper certainly did not want to arrive at—not in his evolutionary epistemology, that is. In his methodology and, more importantly, in his theory of rationality this antagonism between criticism and dogmatism presents the basic assumption.

Further, Popper anchors dogmatic instincts in our nature but does not deal with the question of how does the unique critical disposition come about. He says in passing that "the method of trial and error-elimination is largely based on inborn instincts" (ibid., 25). But this one isolated comment does not tie in with or deal with his more elaborated conception of the inborn dogmatic instinct. But even if we accepted that critical reason is inborn in the sense that it emerges during the evolution of human species we still face the same unresolved problem. If both actors in the cognitive business—instincts and reason—are products of the same evolutionary processes, why do they stand in such an antagonism to each? Shouldn't dogmatism, from the evolutionary point of view, gradually get weaker; and shouldn't criticism get stronger? After all, criticism, and thus reason, has been the main human evolutionary advantage—why hasn't man been gradually transformed into a more "naturally" critical, open-minded, intellectually adventurous being? Or has he?

The way in which Popper approaches this problem is unbalanced and reflects the fact that accommodating criticism and dogmatism side by side is a difficult—if not an impossible—task, especially given his merciless denouncement of dogmatism in both his theory of rationality and his methodology. For Popper, dogmatism is the essence of an irrational, cognitively unproductive and ideologically dangerous attitude that must be eliminated to open the way to critical reason. Yet, as an unintended consequence of his solution of the psychological problem of induction, he granted dogmatism a strong biological (and, during evolution, un-subsiding) status. He would have had to be a magician to reconcile these two powers—not to mention the fact that he did not quite realize the extent of the trouble. He therefore made various moves that try on the one hand to legitimate dogmatism and on the other hand to express the belief in growing critical abilities. These moves are not consistent with each other. Let's have a closer look at them.

Popper has to defend some legitimacy for dogmatism once he anchored dogmatism in our genes. He justifies this step by the argument that our conjectures must be given "fair trial." In order for them to "make their case" they must first "show their mettle" and assert their position with vigour that, for Popper, includes some dogmatic elements. Only then they are "ready" to be targeted by criticism. That is to say, "the critical attitude is not so much opposed to the dogmatic attitude as super-imposed upon it" (Popper 1985, 50).

I think this is a bad move because it corrupts his extreme falsificationism and thus the pillar of his theory of rationality. His concept of negative reason denying the status of rationality to any form of justificationism allows no bargains with dogmatism whatsoever. Popper's philosophy had been driven by *horror dogmatis*; he ascribed to dogmatism the cunning to sneak in a rational discourse in various disguises and undermine it from within. He therefore insisted that extreme measures must be taken to fend off this danger. For him, the only effective anti-dogmatic strategy could be insured by adopting the imperative of falsification, forbidding within a rational discourse any but severely critical assessments of any theoretical claims, aiming at their refutation. This must be totally non-negotiable. Legitimating some positive role of dogmatism jeopardizes the most original—and provocative—aspect of his philosophy and, in my view, devalues his philosophical contribution. How is one supposed to know what amount of dogmatic persistence is beneficial for the "fair trial" of conjectures, and where the limit lies? Accusations of mistrial and demands for another hearing can be used. As Popper stated before—it is either all or nothing since dogmatism can often successfully parade under a critical label.

The second way in which Popper downplayed the clash between criticism and dogmatism was by suggesting that dogmatic instincts can be reformed. He claims that it is possible to suppress dogmatic beliefs by reformulating belief-sentences in objective (criticisable) terms: for example, "instead of speaking of a 'belief,' I speak, say, of a 'statement' or of an 'explanatory theory'" (Popper 1979, 6). Popper applies this procedure to Hume's logical problem of induction and then proceeds to transfer it to Hume's psychological problem of induction, too. He argues: "on the basis of the following *principle of transference*, what is true in logic is true in psychology" (ibid., emphasis in the original).

Following this principle he restates the psychological problem of induction (ibid., 26):

> If we look at a theory critically, from the point of view of sufficient evidence rather than from any pragmatic point of view, do we always have the feeling of complete assurance or certainty of its truth, even with respect to the best-tested theories, such as that the sun rises every day? I think that the answer here is: No.

I do not agree with Popper on this and my answer in this particular case is: Yes. The answer in specific cases would depend on which theory is under consideration, and often the feeling of assurance wouldn't be complete or ever-present. But I question Popper's link between a "critical look" and the "feeling" that he postulates above. Hume was worried about the human predicament, in which sceptical reason undermines beliefs, while our nature compels us to feel and believe regardless of what reason whispers to us. Popper criticizes Hume in other respects but not this one—how can he, out of the blue, claim that reason can modify beliefs—or even overpower them? He does not provide any backing for this far-reaching statement, not to mention the fact that his own grounding of dogmatic beliefs in our genetic hardware (with no such a place given to the critical disposition) rather excludes this possibility.

When he elaborates further on this problem, things become even fuzzier. He classifies the belief that the sun rises every day as a "*pragmatic* belief, something closely connected with [...] our instinctive need for, and expectations of, regularities" (ibid.). So far so good. But this doesn't prevent him from repeating that after we rationally reflect on the available evidence, "we shall have to admit that the sun may not rise tomorrow over London after all" (ibid.). But there is a crucial difference between "admitting" and "feeling" that Popper overlooks in order to make his case. His position is, in my view, wishful thinking that expresses the modern ideal of the advance of reason (i.e. criticism) and its increasing internalisation in our minds and power over instincts.

This corresponds to Popper's thesis that the process of intellectual maturing —either in the individual or the historical sense—can enhance the critical faculties *at the cost of dogmatism*. He seems to assert that a higher degree of intellectual refinement leads to less dogmatism and, conversely, that a lack of intellectual maturity implies stronger dogmatism: "dogmatic thinking, an uncontrolled wish to impose regularities, a manifest pleasure in rites and in repetition as such, are characteristic of primitives and children" (Popper 1985, 49). What Popper probably had in mind was the incapacity of these two groups to produce the world of objective knowledge and to detach ideas from the cognitive agent. But does education necessarily weaken dogmatism? Some "primitives" may be more flexible regarding error-elimination due to the necessity to survive. Children can be open-minded puzzle-solvers. But it is most pretentious to assume that educated people have a higher ratio of criticism to dogmatism. On the contrary, intellectual maturity—or more advanced forms of societies—enables a greater dogmatic sophistication by disguising it in a critical rhetoric. Such "clever" dogmatism is even more dangerous—more difficult to recognize and unmask.

In this context, another interesting area of problems could be investigated concerning social evolution and its link to the evolution of knowledge. As the last "laggard of the Enlightenment" (Popper 1988, 177) he views the development from primitive to modern societies as progress. Modern societies allow public critical discourse, and ensure this right by legal rules, and it is this increasing influence of reason that makes progress, i.e., social order based on democratic principles, possible. But then, the problem arises of whether modern civilization does not pose even bigger dangers to progress, stemming from the impact of various irrational and dogmatic tendencies that can produce even more effective machinery of violence and suffering—more sophisticated forms of totalitarianism or global wars misusing the latest scientific discoveries. Popper addressed this problem forthrightly in his *Open Society* but regrettably did not tie it together with his epistemological investigations.

I think that Popper should have stuck with his original conception of negative reason that assumes an ongoing and unrelenting opposition between criticism and dogmatism, and even makes dogmatism the stronger element. This position explains his extreme falsificationism requesting us to be permanently alert towards any signs of justificationism (both in science and society) entering into the rational discourse. He should have also stated more clearly that against a "fragile" critical capacity of the human mind there stands a hefty justificationist and dogmatic disposition that seems to be resistant to the critical charms. As far as he held this view Popper was consistent with his principal philosophical line. This would be further and consistently reinforced by his concept that dogmatism forms the human *apriori* genetic constitution.

But then, difficulties would arise as how to co-opt this line into the evolutionary epistemology; how to explain why the force of our dogmatic nature that is counterproductive for the growth of knowledge has not subsided during evolution. Two answers can be found in Popper's writings and neither of them is satisfactory. He either suggests that criticism does after all and over time (or through education) win "somewhat" over dogmatism. This answer is in accord with the evolutionary view but contradicts his above-mentioned biological grounding of dogmatism and is presented in a haphazard way without a proper argument. Or, he puts up with the natural power of dogmatism and tries to legitimate its role in evolution. This strategy is lethal for his rationalism and corrupts his most original concept of *ratio negativa*, based on a total exclusion of dogmatism (always teamed-up with justificationism). Popper did not realize the fact that he is alternately exposing or evading one aspect—criticism or dogmatism—jeopardizing thus either his evolutionary theory or his rationality theory. As much as I acknowledge Popper's contribution towards evolutionary epistemology I think he did not face up to and deal with this tension appropriately.

References

Campbell, D. (1974). Evolutionary Epistemology. In *The Philosophy of Karl Popper.* P. A. Schilpp (ed.), La Salle, Ill.: Open Court, 413-463.

Dobzhansky, T. (1970). *Genetics of the evolutionary process.* New York: Columbia University Press.

Grehan, J. F. (1984). Evolution by Law: Croizat's "Orthogeny" and Darwin's "Laws of Growth." *Tuatara* 27 (1), 14-20.

Hahlweg, K. and Hooker, C. A. (eds). 1989. *Issues in Evolutionary Epistemology.* Albany, N.Y.: SUNY Press.

Munz, P. (1985). *Our Knowledge of the Growth of Knowledge: Popper or Wittgenstein?* London and Henley: Routledge.

O'Hear, A. (ed.) (1995). *Karl Popper. Philosophy and Problems.* Cambridge: Cambridge University Press.

_____ (1999). *Beyond Evolution. Human Nature and the Limits of Evolutionary Explanation.* Oxford: Clarendon Press.

Parusniková, Z. (2004). Two Cheers for Karl Popper. In *Karl Popper. Critical Assessments of Leading Philosophers. Volume I.* A. O'Hear (ed.). London: Routledge, 79-99.

Popper, K. R. (1959). *The Logic of Scientific Discovery.* New York: Harper and Row.

_____ (1979 [1972]). *Objective Knowledge.* Oxford: Clarendon Press.

_____ (1973). Evolutionary Epistemology. In *A Pocket Popper.* D. Miller (ed.). Glasgow: Fontana 1983.

_____ (1980 [1945]). *The Open Society and Its Enemies*, vol. II. London and Henley: Routledge.

_____ (1985 [1963]). *Conjectures and Refutations*. London and Henley: Routledge & Kegan Paul.

_____ (1988 [1983]). *Realism and the Aim of Science: Postscript to The Logic of Scientific Discovery*, Volume I. Ed. by W.W. Bartley III. London: Hutchinson.

_____ (2003 [1999]). *All Life is Problem Solving*. London and New York: Routledge.

Schilpp, P. A. (ed.) (1974). *The Philosophy of Karl Popper.* La Salle, Illinois: Open Court.

Radnitzky, G. and Bartley, W. W. III. (eds.) (1987). *Evolutionary Epistemology, Rationality and the Sociology of Knowledge*. La Salle, Illinois: Open Court.

Watkins, J. (1984). *Science and Scepticism.* Princeton: Princeton University Press.

Worrall, J. (1995). "Revolution in Permanence": Popper on Theory-Change in Science. In O'Hear (1995), 75–103.

Wuketits, F. (1990). *Evolutionary Epistemology and Its Implications for Humankind*. Albany, N.Y.: SUNY Press.

10

Reciprocal Containment, Naturalized Epistemology and Metaphysical Realism

Jonathan Knowles

1 Introduction

This chapter is about whether science can provide something like an epistemic justification of itself. In "Epistemology Naturalized," Quine makes the following famous statement:

> The old epistemology aspired to contain, in a sense, natural science; it would construct it somehow from sense data. Epistemology in its new setting, conversely, is contained in natural science, as a chapter of psychology. But the old containment remains valid, too, in its way. We are studying how the human subject of our study posits bodies and projects his physics from his data, and we appreciate that our position in the world is just like his. Our very epistemological enterprise, therefore, and the psychology wherein it is a component chapter, and the whole of natural science wherein psychology is a component book—all of this is our own construction or projection from simulations like those we were meting out to our epistemological subject. Thus there is reciprocal containment, though containment in different senses: epistemology in natural science and natural science in epistemology. (Quine 1969a, 82)

This idea has been widely discussed in the subsequent literature on so-called "naturalized epistemology." One reaction has focused on the idea of epistemology as "a chapter of psychology," something that has led to accusations against Quine of simply changing the subject, since epistemology is an essentially normative discipline (Kim 1988). In response to this, Quine has argued that naturalized epistemology does not jettison the normative dimension of its predecessor, but rather transforms this into "the technology of truth-seeking or, in a more cautiously epistemological term, prediction" (Quine 1986, 664). In other words, epistemology seeks to discover methods that are conducive to the aims of science, using science itself to select and justify these methods.

The general idea of "reciprocal containment" is clearly evident in this idea, but its most explicit application in Quine's writings is in relation to the traditional debate of external world scepticism. Science, says Quine in the quote from

Epistemology Naturalized above, is to tell us how we build up our picture of the world from sensory promptings; hence the containment of epistemology in natural science. But at the same time, this science—that which tells us *inter alia* that this is how we gain knowledge—is part of the very knowledge we thereby acquire, from which it follows that, in being likewise a product of sensory promptings, *it* is "contained" in epistemology. Many, including Quine himself, have taken this situation to lead inexorably to the question of whether we actually can or do know what we think we know in science: whether our sensory promptings and the processes these are subject to in our brains are such that they can vindicate the putative knowledge on the basis of which we (*inter alia*) ask this very question (cf. e.g. Quine 1974, xi ff., 1975). Quine distinguishes himself from the traditional epistemologist, who (allegedly) sees sceptical worries as something to be dealt with prior to engaging in science. But this does not rule them out:

> I am not accusing the skeptic of begging the question; he is quite within his rights in assuming science in order to refute science; this, if carried out, would be a straightforward argument by *reductio ad absurdum*. I am only making the point that skeptical doubts are scientific doubts. (Quine 1975, 288)

This remark has provoked a debate about whether Quine must count himself a sceptical philosopher, insofar as he also cleaves to the doctrines of inscrutability of reference and underdetermination of theory by data (cf. e.g. Stroud 1984, Gibson 1988, Hookway 1988). Put briefly, if Quine sees our scientific theories and postulates as things from which we could rationally demur in favor of some other ontology and world-view given just our sensory input—which seems very much to be his view, right from "Two Dogmas of Empiricism" (Quine 1953)—then it seems we face precisely the *reductio ad absurdum* he sketches in the quote above.

Others disagree with this assessment of Quine's overall position (e.g. Gibson 1988), but I will not be entering this intricate (and partly exegetical) controversy here. Rather, I want to focus on the views of someone who, though very much a Quinean in respect of his reverence for natural science, does not subscribe to the anti-realist and/or sceptical leanings enshrined in the inscrutability and underdetermination theses: Hilary Kornblith. Nevertheless, and interestingly, Kornblith is concerned with the epistemological consequences of naturalism and, not least, with the idea of reciprocal containment: that it seems we need to vindicate our science through science, as much as we do any other part of our knowledge. Kornblith's idea in particular is that a scientifically motivated metaphysics, combined with a scientifically informed account of our cognitive capacities for belief-formation, should mutually support one another and that by doing so will provide a kind of vindication of our overall scientific world-view (cf. especially Kornblith 1993, 1994). Implicit in this is thus at least one central element of Quine's conception of naturalistic epistemology: that it is something that proceeds from within science and is *a posteriori*, but that nevertheless gives no guarantee of success: the possibility of a scientific *reductio* of science remains open in principle. Exactly what this eventuality would lead us into were it to be realized is not something we need actively to contemplate—no doubt it would be a state of considerable epistemic chaos. The point is that its possibility is presupposed in the very idea of what we hope and in some sense believe will instead be the upshot of our enquiries: a scientific vindication of science.

This chapter is concerned to argue against this idea: that is, against the idea that there can be such a scientific vindication of science, or indeed, a refutation thereof (at least assuming we put aside the Quinean line on inscrutability and underdetermination mentioned above). I will not be able to give a watertight demonstration of this, but I do hope to make plausible that it is not, *pace* Kornblith, a project which concrete findings in the various sciences, particularly empirical psychological studies of reasoning, can make a significant contribution to—at least without further, questionable assumptions. I will be doing this against a broadly naturalistic background—one on which Quine's notion of reciprocal containment in some form or other applies. Thus I will not be arguing against Kornblith's particular claims or naturalism, but rather questioning their epistemological significance—their capacity to vindicate science.

In the following section, I will sketch in more detail how Kornblith conceives his combined metaphysical and epistemological project. In sections 3 and 4 I then present, respectively, two horns of a dilemma for the line that this project can provide a vindication of science: given one understanding of the project its epistemological significance turns out to be at best very small—and plausibly not what Kornblith had in mind—whereas on another understanding, though the significance might have been larger, the understanding turns out in fact to be incoherent and is thus unavailable. In both sections the underlying commitment to reciprocal containment plays a crucial role in the dialectic. I end with a brief conclusion concerning the prospects for naturalized epistemology more generally.

2 "Naturalism: both Metaphysical and Epistemological"

The title of this section is that of Kornblith's 1994 article referred to above. (The epistemological components of the view are more thoroughly presented in the 1993 monograph *Inductive Inference and Its Natural Ground*, but I will focus on the paper here, and all future otherwise uncited page references are to this.) I will outline Kornblith's view as presented in this article to show more precisely how he thinks science may provide a kind of vindication of itself.

Let us start with the metaphysics. Kornblith tells us that his concern is to articulate the metaphysical view of nature implicit in contemporary natural science: a scientific metaphysics. (If one asks why our metaphysics must be that of science, the answer is that science has proved itself our most successful, and hence by implication, most reliable source of information about the world around us. I will not here be questioning the possible "scientism" of this view, nor the assumption that science needs a metaphysics.)[1] A proper respect for science in metaphysical theorizing leads to the following theses (Kornblith 1994, 44):

(1) Anti-reductionism: the predicates of higher level sciences do not pick out properties that reduce to those of the basic science, considered either as types or as tokens.

(2) Materialism: everything is physically constituted (i.e. Cartesian dualism is false).

(3) Causal powers operate at the level of higher sciences as well as basic sciences.

(4) Natural kinds exist as homeostatic property clusters, determining their peculiar forms of causal interaction.

(5) Causation and laws have to be understood in a non-regularity, i.e. non-Humean fashion.

Many of these are interconnected. Following Kornblith's discussion, the idea I will focus on is (4), that of natural kinds as homeostatic property clusters—i.e. properties that co-vary with one another and are causally dependent on one another in such a way that they persist over time as such a cluster, and not merely by chance. The motivation and justification of this account need again not detain us (see Boyd 1991), since it is the use Kornblith makes of it in his epistemology that will be important.

So much for metaphysics—what about epistemology? Kornblith tells us that "[a] proper epistemological theory should explain how knowledge is possible" (Kornblith 1994, 44), and goes on to outline a more or less Quinean view of how this should be achieved, namely, not *a priori*, but through science itself. That is, assuming science as an object of knowledge for us, can science also tell us how it could be such an object? The important point for us—to be critiqued—is that the traditional aim of showing how knowledge is possible *can* still be pursued, just by non-traditional means. The idea, roughly, is that we have two mutually constraining bodies of information. On the one hand, we have our scientific theories from physics, chemistry, biology and so on, which we take to be paradigms of knowledge, and which yield the metaphysical world view outlined

---

[1] Both can of course be questioned. For discussion aiming at vindicating something like the hegemony of natural science (with respect to e.g. human science), see Knowles (2008a). For critical discussion of much contemporary metaphysics in analytic philosophy, see Ladyman & Ross *et al* (2007), and for an argument against the need for a metaphysics for science at all, see Knowles (2008b).

above. On the other hand, we can pursue psychological investigations of our natural cognitive capacities. These two sets of data will apply mutual pressure to each other in such a way that we can come to understand how creatures with capacities like us could know things of the kind we seem to (including theories about our cognitive capacities). Thus, just as Darwinian theory united and thereby expedited progress in ecology, microbiology and ecology, so the kind of enquiry Kornblith outlines aims to brings together the epistemology and metaphysics of science in an attempt to show how knowledge is possible, and thus give ever greater credence to our overall scientific world-view (ibid., 45).

What is perhaps most distinctive of Kornblith's approach is that he tries to do this by reference to real cognitive science, rather than armchair speculation. Quine himself in "Natural Kinds" avers that thanks to Darwinian processes of natural selection, "creatures inveterately wrong in their inductions have a pathetic but praiseworthy tendency to die out before reproducing their kind" (Quine 1969b, 126). Since we have not died out, this is meant to give us some—scientific—reason to think that our inductively based science could be an object of knowledge for us. The argument is hardly watertight and is in any case highly abstract. Kornblith (1993, ch. 1) mentions it to set it aside; he aims instead to give some more specific evidence that creatures like us might have latched onto the kind of world science seems to reveal to us.

The first key aspect of Kornblith's naturalistic epistemology is its stress on the ecological situatedness of our knowledge-gaining capacities. The paradigm here is our perceptual capacities. In accord with recent thinking in cognitive science, argues Kornblith, a perceptual capacity should be viewed as finely tuned to the specific physical and environmental conditions of its realisation in us: an environment with very different physical background conditions, or in which the objects we perceive obeyed substantially different regularities would not be one we could derive reliable information about given the perceptual faculties we in fact possess. This is precisely what e.g. visual illusions demonstrate. Kornblith contends, in the spirit of an evolutionary approach to psychology generally, that we must view all our cognitive faculties, including those involved in inference and reasoning, as similarly keyed to the precise environments they have evolved to cope with.

This perspective is then applied to some widely discussed data from experiments on reasoning, in particular that of Tversky & Kahneman suggesting that humans are naturally inclined to obey the "law of small numbers" (cf. Tversky & Kahneman 1982)—i.e. to draw inductive inferences on the basis of small and—from a statistical point of view—unrepresentative samples. Kornblith (1993, ch. 5) sketches the kind of threat this and other experiments can seem to pose to the possibility knowledge: if they show us that we are inherently bad reasoners, then surely we cannot be sure that the generalizations of science we depend on—*inter alia* in expressing the results of reasoning experiments—are to be trusted. We seem to face, not perhaps a full-blow Quinean refutation of science by science, but at least a large internal problem.

Kornblith however does not think the experiments show us that we are bad reasoners, for they neglect precisely the biological context of application of our reasoning capacities. Tversky & Kahneman and many others who have assessed human reasoning characteristically assume a domain-general model of the principles, of the kind standardly assumed in classical treatments of deductive and inductive logic. But this is not an appropriate model for understanding real human reasoning, which is keyed to contingent features of our standard environment (apart from the general evolutionary orientation, this view is also supported by the unfeasibility of domain-general reasoning principles for finite beings like us—cf. Cherniak 1986). Assuming an alternative, domain-specific model, and given moreover the metaphysical picture Kornblith sketches, we can come to see that obeying the law of small numbers will in fact in many cases be a reliable mode of inference, at least to the extent we correctly identify natural kinds and use predicates for these kinds and their associated properties in our inductive reasoning. For if we have in this way "latched onto" natural kinds and their properties, then even a single observation of such a property will allow us reliably to project this to further cases (e.g. given water is a natural kind and that fluidity at room temperature is one of the homeostatic cluster of properties that constitutes it, then just one observation of water's fluidity will put us in a position reliably to project this property to all further instances of water in the same kinds of conditions). Kornblith argues further that we indeed are natively disposed to "latch onto" natural kinds viewed as homeostatic property clusters: this requires sensitivity to property covariation, and though some work exists suggesting we are bad at this, it turns out that when the number of properties involved is relatively high (at least greater than two), our ability to detect covariation is quite high (cf. Kornblith 1994, 47 and references).

This gives some reason for optimism, but Kornblith admits that our capacities are not perfectly tuned to all the different kinds of reasoning scientists might engage in. Thus it is important that we can also learn to be more "rational," i.e. better reasoners. But again this is not to be treated with a mere wave of the hand, rather we need to understand the nature and scope of our learning mechanisms, in relation to their proper biological contexts. Ultimately this line of thought will lead us, argues Kornblith, to a view of knowledge as a natural kind, at least insofar as a genuinely scientific epistemology is possible (a view more fully developed in Kornblith 2003).

Kornblith is thus clearly optimistic with respect to naturalized epistemology, even though this optimism is not (supposed to be) based on any *a priori* argument but only *a posteriori* scientific investigation combined with an appropriate metaphysical view of science's achievements. There is an awful lot that might be and has been said about Kornblith's view as whole, and about the particular aspects of it reviewed here. What I want do here however is to a large extent abstract from the details and focus instead on the general Quinean idea that it exemplifies: the idea of a scientific vindication of science. Is such a thing really possible–that is, in what way do the kinds of studies Kornblith outlines really have epistemological significance? Though I will still be concentrating on Kornblith, for the sake of concreteness and thus clarity, I hope that the considerations I will adduce can have application beyond the relatively narrow confines that Kornblith's particular constellation of views defines.

3 The First Horn of the Dilemma

I want to start with a reflection that some might see as an objection to the line of thought just presented; I think as such it is not a good one, but it does function as a way into thinking about the kind of epistemological significance Kornblith's project can actually have. Kornblith argues, firstly, that our scientific metaphysics fundamentally involves the positing of natural kinds (understood in a certain way); secondly that, if these natural kinds are such that we in some way "hook up" to them in our inductive reasoning, then our innate reasoning tendency to obey the law of small numbers will not lead us astray; and finally that we indeed are likely to "hook up" to them in virtue of our ability to detect covariation amongst larger numbers of covarying traits. Thus we have a story about how the world is and a story about how we could know such a world, and hence our faith in the story encompassing both increases. But of course this is not the *whole* story about our knowledge of the world. For example, Kornblith relies on an account of what scientific knowledge we have as the basis of his inference that there are natural kinds understood as homeostatic property clusters. It seems clear therefore that we need in principle an epistemology of that prior account: if we need an epistemology for an account A in order in the final analysis to be able to accept A, and if A depends on a further account A', then surely we need an epistemology for A' too, at least in principle. Similarly, Kornblith relies on the general knowledge of psychological experimentation that underlies the studies of reasoning and covariance detection, our knowledge of which we would again in principle need an account of. Furthermore, these accounts would in turn have presuppositions we would have to demand an epistemology for, and so on and so forth *ad infinitum*. Somewhat similarly, the account of learning will presuppose other items of knowledge that we will also have to give an epistemology of–not for the sake of the learner, but for the scientist or philosopher who wants to understand how we could learn to be more rational and scientific.

The point can be presented more schematically as follows: if we are concerned to provide a naturalistic epistemology (in the sense presently under discussion), then we need, by reciprocal containment, to understand how the world as viewed by science (let us symbolize this as "X") leads us (something like the collective scientific subject, which we may denote "S") to know X–this being also a scientific truth about how external states of affairs impact on S. But further, since by naturalism S will know *these* facts ultimately on the basis of her contact with the world (observationally, phylogenetically or ontogenetically), we must also have an account how X leads S to know how X leads S to know X; and again, by the same reasoning, how X leads S to know how X leads S to know how X leads S to know X. And so on, *ad infinitum*–given reciprocal containment.[2]

As indicated above, if someone thought this was in itself a trenchant objection to Kornblith's naturalized epistemology, I think they would be mistaken. The reason is just that the kind of regress just adumbrated can be seen as *virtuous*, each step leading to a new avenue for investigation, which in turn opens up new avenues again. So long as any particular step is such that we find no reason to deny we can have knowledge of *p* at that step–whatever *p* is–then everything would seem to be in order. Moreover, insofar as each new step presupposes the previous one, each new vindication would seem to increase the overall coherence of our body of belief–we can consistently add another belief to the class of beliefs we already have.

We have now sketched what I take Kornblith would want to see as a virtuous epistemic process: at each new step, it is in principle left open that the attempt to vindicate our knowledge of *p* might fail, and concomitantly, each time we successfully demonstrate knowledge is in fact possible for *p*, the overall coherence of our total belief system increases. I think however we need to consider more carefully whether or in what sense any of this might be seen as a vindication of science. Two things need to be noted.

To begin with, when we talk of demonstrating our ability to know *p* for a given *p*, it is important to be clear that what we really mean is demonstrating that there is as things stand no reason to think we can not know *p*. It must be essentially a negative result we aim at, not a positive one, since, as we have seen, any positive pronouncement depends on further, indefinitely many other levels of knowledge whose veracity in turn await vindication. What this means is that the epistemological significance of such "demonstrations" cannot go beyond

---

[2] I should point out straightaway that I do not mean by this to imply that an individual to be personally justified in believing some proposition, *p*, must know all the conditions for the truth of *p*–which would be absurd (at least to the extent there is a useful notion of "personal justification"). This piece is not concerned with justification in this sense, but rather with the idea of the justification or vindication of *science* as an object for human knowledge.

the epistemological significance any putative demonstration to the contrary would have—demonstrations that we cannot know that *p*.

This means that we can inquire into the epistemological significance of the positive demonstrations in question by asking how significant the opposing, negative demonstrations would be. Here however we need to note a second point: in reality it seems highly unlikely that these would be "demonstrations" in anything like the true sense of the word. The idea that we might be able to show beyond all possible doubt that we *cannot* know *p* for any given *p* from empirical data about our reasoning abilities seems, to put it bluntly, more or less unthinkable (remember that we are not now considering general arguments such as might stem from Quine's underdetermination or inscrutability theses). Nor does it seem likely that the data in question would suggest anything like overwhelming reason to believe we cannot know *p* for any *p*. Rather—as in Kornblith's own real-life examples—the plausible scenario is one in which certain results from psychology present an *apparent challenge* to our claims to know the world that we can try to rebut by seeing them in relation to the world they concern and our biological context more generally.

Staying with these challenges for the moment, we may now ask exactly what the epistemological significance of such data from psychology really can have. If naturalized epistemology is going to have the significance Kornblith wants it to have, these challenges should lead us, in some significant degree, to doubt our scientific knowledge. However, it seems very clear that this is not our reaction in the case he discusses; nor is it at all clear that such evidence ever would or could lead to such a reaction. The discussion of Tversky & Kahneman's results above seems to make the first point clear: that we obey the law of small numbers never led anyone to doubt our science, even before we started looking at the tendency in its proper biological context. Kornblith might protest that it *should* lead us to do this, at least to some extent. However, it is unclear on what grounds. If we have a large and evolving body of belief that is highly explanatory and coherent, the discovery of some new data that do not automatically fit in with what we already think we know is hardly reason to doubt the body of knowledge as a whole. What we have seems at best something more like what Kuhn (1970) calls an *anomaly*—a result that at the time in question does not clearly fit into our going "paradigm," but which there is, considered in relation to the mass of coherent knowledge we otherwise take ourselves to have, no reason to think is any kind of threat to this.

At this point, it might be claimed that if one were to find—and fail to resolve—several such "anomalies," then faith in our overall belief system would begin to waver. However, it is hard to see that however large the number of problems our attempts ran into, we would be inclined to give up our scientific belief-system as whole. After all, as long as most of it remains internally coherent, surely all is well. Moreover, there is no question of replacing it wholesale with something else (as Kuhn suggests has happened at a smaller scale during the history of the particular sciences).[3] Given that our physics, chemistry and biology all remain internally consistent, plus consistent with one another, and that all of this is consistent with pretty much everything else we know about the natural world from science, including psychology—except perhaps the bits that Kornblith details, concerning certain aspects of our reasoning—then surely we would not rationally be obliged to give this all up. Nor is it clear from what perspective we could really see it as in any way threatened.

It would seem to follow, *mutatis mutandis*, that finding out that things are in fact "OK" for any or even many levels is similarly epistemologically of no great moment. There was no real threat to our knowledge in the first place. Rebutting the "challenge" is an interesting theoretical exercise but not a contribution to epistemology in the sense under consideration: a vindication of science by science.

One might at this juncture argue the kind of project Kornblith sketches nevertheless increases the overall coherence of our body of beliefs, by adding more beliefs to it. But does that justify anything? I think one should be a little wary of simply assuming a coherentist theory of justification in any substantial sense: my arguments so far have not presupposed such a theory; moreover, many, naturalists and non-naturalists alike, would oppose it (see e.g. Williams 2001, ch. 11 for a general critique of coherentism). I also take it, what seems to be commonly accepted, that commitment to Quine's largely negative idea of a web of belief does not in and of itself commit one to coherentism *qua* substantive epistemological theory of justification.

That is one point. But let us for the sake of argument assume that something at least like coherentism is viable, at least to some degree—or, what for us amounts to the same thing here, that Quine's web of belief idea does after all involve something like coherentism. Let us now ask if the accretion of more and more items of knowledge—in particular about the ecological reliability of our belief-forming capacities—would lead to greater overall justification of this web. Here I would argue that though the answer must be yes, the epistemological significance is not great. In particular it is no greater than the epistemological significance of adding new knowledge of any other kind. New findings in physics, chemistry, biology and so on would contribute just as much to the "vindication" of science as the kinds of results and ideas Kornblith draws attention to. I take it this amounts to saying that it is not this kind of vindication of science Kornblith is aiming at.

---

[3] I do not assume Kuhn is right about this, but merely point out that it has in any case no obvious implication for our discussion here.

Of course, to be no great moment is not to lack significance entirely. But we must not forget what is meant to be at stake here. According Kornblith's (and many others') view of naturalized epistemology, the kinds of psychological studies and theoretical considerations he draws attention to are meant to have a peculiarly central role in providing an overall vindication of our scientific world-view. We seem instead to have concluded that they are just a very tiny part of an enormous puzzle. Psychology, or psychology of a certain special kind, has no special role to play in the vindicatory project—if indeed we now want to call it that at all.

4 The Second Horn of the Dilemma

We may tend to feel that things must be otherwise—that the kind of *a posteriori* project Kornblith sketches is in fact of quite some epistemological moment. Why might we feel this? Perhaps we can begin to find out by asking what it would take for findings from reasoning psychology (as I will from now on simplifyingly put it) to have a crucial epistemological role to play in the picture of justification we have so far been operating with.

So: assuming the web of belief model and (for the sake of argument) some minimal epistemological coherentism as part of this, what would it take for reasoning psychology to have the significance Kornblith clearly sees it as having? The only viable answer would seem to be: lying close to the centre of the web, rather like logic or mathematics—or, at least, relating to an idea that lies close to this centre. These central ideas, not part of but a background for the sciences, though also in principle on a continuum with them, would seem to be such that giving them up would initiate an enormous upheaval in our most basic epistemic practices. I won't here be considering the very idea of such centrality and its epistemological role, but will instead be focusing on reasoning psychology in relation to such a central core. At first blush, it is hard to see why it might have such a central role, or be related to something that does. Why should empirical facts about reasoning abilities have some kind of fundamental status in a way that—say—general relativity does not?

I think nevertheless that this diagnosis of our intuitions is on the right lines. Reasoning psychology is in particular often seen as embedded within a certain picture, and it is this picture that I think for many lies near the heart of the web. The picture is in effect a metaphysical conception of the relation between mind and world: the view of the world as given on the one hand, and the mind as knowing it by virtue of representing it on the other: the picture today often referred to as "metaphysical realism."

Now it may seem odd for a naturalist to have a so fundamentally metaphysical picture at the centre of her web of belief, but let that pass (perhaps it is odd for a naturalist to have anything there when it comes to that). For the moment, what is important is that if we assume this metaphysical picture, then it might seem that the kind of empirical results Kornblith describes and envisages could indeed by problematic for—or alternatively, vindicatory of—science. Why? Because the issue would concern nothing less than a potential breakdown in a very significant part of the world as a whole: the mind/world interface. Given there is such an interface, finding out something that suggested, say, an incongruence between what is on the one side and what is on the other would be a highly significant result. It wouldn't matter that the amount of knowledge in question was very small compared to what we otherwise take ourselves to know: it could be crucial to the whole picture hanging together, so to speak. The difference is like that between a carpet that has a few loose threads, and a piece of clothing having a few loose threads at some significant join.

Let us (for reasons that will soon become apparent) underline this idea with a picture. In the following, "X" represents the world, and the circle with the "X" inside the knowing subject representing the world: the line is there to emphasize the "interfacial" nature of the relationship:

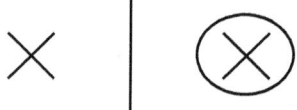

Now Kornblith is as far as I am aware a kind of metaphysical realist—indeed, his natural kind view of knowledge seems highly integrated with this overall commitment (cf. his 2003). Can Kornblith then, in response to the critique of the previous section, now say that he does indeed assume this metaphysical picture as a central component of the web of belief, and that that is precisely why studies in reasoning psychology have such a central place in epistemology? I think he might well want to say this; moreover, I think he would have to say it in order to ground the epistemological significance of his project. Here however the naturalistic commitment to reciprocal containment becomes more than merely a problem of regress.

In considering Kornblith's project in section 3 we arrived at the idea that there is no end to the levels of knowledge about which one would in principle have to ask about the epistemological possibility of: for each new one, there would be a question as to how one could know that. That kind of reciprocal containment was—rightly, I argued—not seen as raising a principled problem for naturalised epistemology. However, we also saw that the project as understood in section 3 failed in and of itself to generate any real epistemological significance for the scientific findings Kornblith adumbrates. What I am now

Reciprocal Containment, Naturalized Epistemology and Metaphysical Realism   137

suggesting is needed in addition is a metaphysical picture of knowledge: a view on which this involves us representing reality to ourselves.

The problem though is that reciprocal containment cannot in fact make sense when we consider this kind of representational picture. The idea would seem to have to be that the world contains our knowledge, whilst this in turn "contains," i.e. represents, the world. But this kind of mutual containment is, interpreted in this metaphysical realist way, just incoherent: one thing cannot contain another that in its turn contains the first thing. To make this absolutely clear, we should first revise the above picture of metaphysical realism to fit with more naturalistic assumptions—i.e. put the subject clearly in the world:

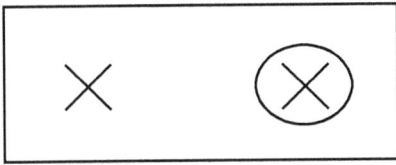

Of course, as we have already noted, things could never be quite that simple: no one *person* knows or could know all of science. The picture above is nevertheless an idealization—the "subject" seen as some kind of collective knowing subject of modern science, perhaps—that the current interpretation of reciprocal containment must make room for if it is to play the epistemological role envisaged for it by naturalized epistemology. All the same, a moment's reflection should convince that it does not give a correct rendering of reciprocal containment: what the subject knows must be *everything* that is (taken to be) the case, *including facts about how her knowledge is related to the world*, i.e. that it is a part of it. This might suggest that need for something more along the following lines (for presentational purposes, the subject has now become an oval!):

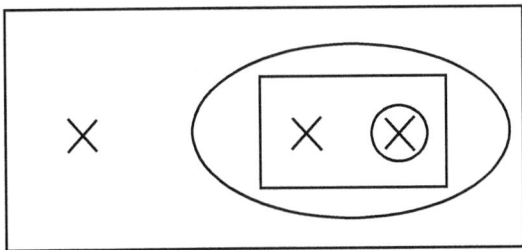

But of course this attempt to revise the first picture is futile—what is known or represented is still not the whole picture—and the same will apply to any future attempt at refinement. We have failed to give our subject the knowledge that she must have for reciprocal containment to hold, and this will necessarily be the case however we adjust the picture.

I should stress that though I have used pictures here to make my point, it does not in fact depend in any way on any kind of pictorial notion of representation. The point is that metaphysical realism holds that the world is represented in the subject, and the subject with her representation is in the world. And this is incoherent.

In sum, one cannot make sense of this metaphysically realist interpretation of reciprocal containment. But if this cannot be made sense of, then—given also what was argued in section 3—there is little to naturalized epistemology conceived as the project of attempting to vindicate our scientific picture of reality—a reality that we might conceivably be cut off from. Just again to be absolutely clear, I am not claiming that we (individuals) cannot be brains in a vat, woefully out of touch with reality. My point is that *science* itself doesn't furnish us with that kind of picture of our epistemic situation, together with a strategy for arguing that scepticism is not in fact warranted. We cannot so much as frame the problem in this way for science as a whole. In effect, what we have shown is that metaphysical realism leads to incoherence when combined with reciprocal containment: to the extent that one seeks to understand naturalized epistemology through the latter idea, one cannot make use of the former. But without the former, "naturalized epistemology" gives us very little of epistemological significance.

5 Conclusion

Declarations of naturalism in epistemology have it seems to me a rather ambiguous import. On the one hand, there are naturalists that accuse traditionalists of missing important epistemological points by refusing to attend to empirical data of various kinds, and generally "de-psychologizing" epistemology (for an overview, see Kitcher 1992). This is a metadebate, but is meant to have clear consequences for epistemology as practised. On the other hand, there are naturalists who see no point in epistemology *per se* whatsoever—or rather, less scathingly, that it is a praiseworthy but forlorn aspect of "first philosophy." My view is that the latter take is more plausible in relation to many different aspects of the "naturalised epistemology" debate. In Knowles (2003) I tried to show this with respect to the project of deriving norms for scientific enquiry, arguing that this is unnecessary for optimal rational belief formation, on a naturalistic view of what this involves. In Knowles (in preparation) I argue in a somewhat similar spirit there is no naturalistic notion of peculiarly *epistemic* normativity. I see the present chapter as plying a similar overall line with respect to the traditional external world debate: the project of vindicating science through science is otiose. Of course, I have restricted myself here to a view that stresses actual work in the sciences, rather than abstract arguments of the kind Quine uses. Showing what significance these arguments and ideas might have is a task for future consideration.[4]

References

Boyd, R. (1991). Realism, Anti-Foundationalism and the Enthusiasm for Natural Kinds. *Philosophical Studies* 61: 127-148.
Cherniak, C. (1986). *Minimal Rationality*. Cambridge, MA: MIT Press.
Gibson, R. (1988). *Enlightened Empiricism: An Examination of W.V. Quine's Theory of Knowledge*. Tampa, Fl.: University of South Florida Press.
Tversky, A., & Kahneman, D. (1982). Belief in the Law of Small Numbers. In *Judgment under uncertainty: Heuristics and Biases*. D. Kahneman, P. Slovic and A. Tversky (eds.). Cambridge, Cambridge University Press.
Kim, J. (1988). What Is "Naturalized Epistemology"? In *Naturalizing Epistemology*. 2nd Ed. H. Kornblith (ed.). Cambridge, Mass.: MIT Press.
Kitcher, P. (1992). The Naturalists Return. *Philosophical Review* 101: 53-114.
Knowles, J. (2003). *Norms, Naturalism and Epistemology: The Case for Science without Norms*. London: Palgrave.
_____ (2008a). Is Naturalism a Threat to Metaphysics? *Norsk filosofisk tidsskrift*, 43: 23-32.
_____ (2008b). Two Kinds of Non-Scientific Naturalism. In *Beyond Description: Normativity in Naturalised Philosophy*. M. Milkowski and K. Talmont-Kaminski (eds.). College Publications (in press).
_____ (in preparation). Naturalism and Epistemic Normativity.
Kornblith, H (1993). *Inductive Inference and its Natural Ground*. Cambridge, Mass.: The MIT Press.
_____ (1994). Naturalism: Both Metaphysical and Epistemological. In *Midwest Studies in Philosophy*. Vol. 19: *Philosophical Naturalism*. P. French, T. Uehling and H. Wettstein (eds.). Notre Dame, Ind.: University of Notre Dame Press, 39-52.
_____ (2003). *Knowledge and its Place in Nature*. Oxford: Oxford University Press.
Kuhn, T. (1970). *The Structure of Scientific Revolutions*. 2nd Ed. Chicago: The University of Chicago Press.
Ladyman, J. and Ross, D. (with D. Spurret and J. Collier) (2007). *Everything Must Go: Metaphysics Naturalised*. Oxford: Oxford University Press.
Quine, W. V. O. (1953). Two Dogmas of Empiricism. In *From a Logical Point of View*. Cambridge, Mass.: Harvard University Press.

---

[4] Material from this chapter has been presented at various workshops and seminars in Trondheim and Kazimierz during the past two years. Thanks to the participants for feedback, especially Hilary Kornblith for making me see (as I hope I now do!) the dialectical situation more clearly.

_____ (1969a). Epistemology Naturalised. In *Ontological Relativity and Other Essays*. New York: Columbia University Press.

_____ (1969b). Natural Kinds. In *Ontological Relativity and Other Essays*. New York: Columbia University Press.

_____ (1974). *The Roots of Reference*. LaSalle: Open Court Press.

_____ (1975). The Nature of Natural Knowledge. In *Quintessence: Basic Readings from the Philosophy of W.V. Quine*. R. Gibson (ed.). Cambridge, Mass.: Harvard University Press.

_____ (1986a). Reply to Morton White. In *The Philosophy of W.V. Quine*. L. Hahn and P. Schilpp (eds.). LaSalle, Ill.: Open Court Press.

Stroud, B. (1984). *The Significance of Philosophical Scepticism*. Oxford: Oxford University Press.

Hookway, C. (1988). *Quine*. Cambridge: Polity Press.

Williams, M. (2001). *Problems of Knowledge*. Oxford: Oxford University Press.

11

Evolutionary Ruminations on "the Value of Knowledge Intuition"[1]

Christos Kyriacou

1 Introduction

Recent debates in epistemology have paid considerable attention to the so-called epistemic value problems.[2] That is, problems that beset the value of knowledge. These epistemic value problems grow out of the pervasive intuition that knowledge is in some robust sense good (or "valuable"). In what "robust" sense is knowledge considered valuable is to be explained in section II and until then we can rely on our fairly intuitive grasp of the notion. For the time being, let us call this pervasive intuition "*the value of knowledge intuition*," or VKI for short.

We are all well acquainted with VKI because in our everyday life we all very often find ourselves valuing knowledge. That is, we find ourselves considering instances of knowledge to be good (in some sense). Examples are abundant and wide-ranging from the more practical instances of everyday life (e.g. how to use a blender) to the highly theoretical knowledge of mathematics, philosophy and special sciences (e.g. the solution to Fermat's last theorem, the semantic paradoxes or the Big Bang theory).

Indeed, VKI is so pervasive that some might feel inclined to infer that is not simply an intuition but an intuition universally entertained. And given that there is no reason to think that such a universally entertained intuition is deceptive, they might conclude that knowledge is beyond any reasonable doubt to be considered valuable. Really, they might contend, there is not much logical space for taking knowledge to be valueless or, even worse, positively evil.

But this conclusion is a bit too quick as anti-epistemic views are sometimes being expressed. One can find people (laymen and academics) voicing such anti-epistemic views. Sometimes you hear people asserting that knowledge is

---

[1] I would like to thank Matthew Chrisman for helpful comments.
[2] See Zagzebski (1996), Williamson (2000), Kvanvig (2003), Pritchard, Millar and Haddoch (2010). Questions of epistemic value go back to Plato's *Meno* where the question whether knowledge is more valuable than true belief is being discussed.

the royal road to depression and misery, that ignorance is bliss etc.³ Such anti-epistemic views, namely, views that question the reliability of VKI (and perhaps even its universal entertainment) and consider knowledge to be positively evil can, arguably, be found in works of literature, theology and even philosophy.⁴

Yet, although there are things to be said about such anti-epistemic views this is not the right place for this task. I will set aside such sceptical views about VKI and, subsequently, the value of knowledge itself and assume that they are mistaken. Thus, sceptics that do not find VKI compelling can stop here; or at least read it out of intellectual curiosity. The rest of the essay assumes that the pervasive intuition that knowledge is valuable is a reliable one. It correctly tracks that knowledge is in some robust sense valuable.

Scepticism about VKI and the value of knowledge itself set aside, a key epistemic value problem that grows out of VKI is the so-called "*core value problem*," or CVP. In the light of VKI, CVP asks why knowledge is valuable at all. Here I concern myself only with VKI that lurks in the background and essentially motivates CVP. The rest of the epistemic value problems growing out of VKI won't show up in the ensuing discussion, as we are not here interested in addressing them.⁵

What we are here interested in addressing is the question whether evolutionary psychological considerations could potentially inform why knowledge is pervasively found to be valuable. That is, why we seem so naturally and effortlessly disposed to entertain VKI. The obvious methodological question then is how such an evolutionary psychological explanation of VKI should go.

The first thing to be noticed is that there seems to be a *psychological mechanism* operating behind VKI and any psychological explanation of VKI should go through an explanation of the origins and function of this psychological mechanism. There is a stable, perhaps innate, psychological mechanism that disposes agents to find knowledge valuable. If this diagnosis is right, then evolutionary psychological considerations should elucidate the origins and function of this mechanism, if they are to account for the pervasive VKI. But let me first explain the reasons that ground the diagnosis that there is a psychological mechanism operating behind VKI.

Very briefly, the first reason that inclines me to think that there is a psychological mechanism in operation behind VKI is its *pervasive*, if not universal,

---

³ Compare Plato's example in *Philebus* (12d1-3) of "the fool who is full of foolish opinions and hopes and is pleased."
⁴ Views of an anti-epistemic ilk are sometimes attributed to Romanticists like J.-J. Rousseau and other anti-rationalists like Schopenhauer and Nietzsche. But as I am far from being an expert in their work this claim should be treated with some caution.
⁵ A nice unpacking of the epistemic value problems growing out of VKI can be found in Pritchard, Millar and Haddoch (2010, ch. 1)

nature. Few people, I think, will dare to question the reality of VKI and these are bound to have some hard time defending their view. They are bound to have such a hard time because even sceptics about the reality of VKI will be forced to accept that in their non-philosophical moments of everyday life they find knowledge valuable. Our psychological mechanism goes on finding knowledge valuable even in the case of sceptics who would like to claim that VKI is not a real one. This Humean point leads to a second reason.[6]

The second reason is that VKI seems to be almost involuntary. We seem to *find* ourselves valuing instances of knowledge as if this is something we do all the time but we only consciously realize it at our more reflective, philosophical moments. Some might want to go even further here and based on these reasons talk in terms of an innate "module" operating behind VKI but I need not pursue these considerations.[7] All I need to suggest for present purposes is that there is some sort of psychological mechanism operating behind VKI.

That being said, my goal in this chapter is to provide some tentative and exploratory evolutionary ruminations that could inform our understanding of the psychological mechanism operating behind our disposition of finding knowledge valuable. Such evolutionary ruminations are intriguing because by informing our understanding of the psychological mechanism operating behind our disposition of finding knowledge valuable, they could provide us with an evolutionary psychological explanation of why we do have this pervasive VKI. In other words, explain why knowledge pervasively strikes us as valuable.

With this much by way of introduction, we can now sketch how the discussion will unravel. First, in section 2, I introduce CVP which grows out of VKI and explain in what "robust" sense knowledge is found to be valuable. As I explain, knowledge is found to be valuable either instrumentally or finally. In other words, it is found to be good either as a means to action or for its own sake. I call instrumental value *practical* and final value *pure epistemic value*.

Afterwards I explain that evolutionary considerations, if they are to adequately account for VKI, should be able to explain why we entertain VKI in regard to both the practical sense and the pure epistemic value sense. That is, why we seem naturally disposed to think of knowledge as practically valuable or as purely epistemically valuable. I take this to be a *desideratum* for any adequate explanation of the psychological mechanism operating behind VKI.

---

[6] This point is Humean in character because it parallels Hume's (1985) famous point about inductive scepticism. After giving impetus to scepticism about inductive inferences, Hume argued that as human agents (with the psychology we have) we are bound to go on relying on inductive inferences, even if we think they are irrational.

[7] Talk of modules stems from the work of Fodor (1983) and Tooby and Cosmides (1992), though note that Fodor does not rely on evolutionary considerations as Tooby and Cosmides (1992) do.

Second, in section 3, I explore how evolutionary considerations could potentially inform our understanding of the origins and function of the psychological mechanism operating behind VKI, that is, our disposition to finding knowledge valuable. Through exploring the origins and function of the psychological mechanism operating behind VKI, I ruminate on how evolutionary considerations could explain why we find knowledge valuable (instrumentally or finally). In the end, in section 4, I review and summarize the argument.

2 VKI and a Desideratum

Epistemic value problems like CVP (or the so-called *Meno* problem, namely, why is knowledge more valuable than true belief) are widely accepted to constrain any plausible theory of knowledge (See Zagzebski 1996, Kvanvig 2003, Pritchard, Millar and Haddoch 2010). Any plausible theory of knowledge must not only provide an account of the nature of knowledge but also must address problems of epistemic value, if it is to have any luck in the dialectical marketplace.

Such a theory of knowledge can address the problems either by *vindicating* the intuitions that lurk in the background of the epistemic value problems or by *explaining them away* in an adequate manner. But no matter how these problems are to be addressed, a theory of knowledge that does not address the epistemic value problems is at best to be considered incomplete. This much goes uncontested in epistemological circles and it is widely accepted as a legitimate dialectical constraint on any plausible theory of knowledge.

An epistemic value problem is the already introduced CVP, a problem that lies at the core of the debate about epistemic value because it asks the fundamental question of why knowledge is valuable at all. Intuitively, we find knowledge to be valuable and CVP exactly asks why knowledge is valuable at all. As can be easily grasped, CVP is a direct product of VKI because it grows out of the pervasive intuition that knowledge is good or valuable. The pervasive VKI naturally gives rise to the question of why is knowledge to be considered valuable at all.

Now, as it is often noted in the literature, we seem to find knowledge valuable in two senses (Pritchard, Millar and Haddoch 2010, ch. 1). Like other sorts of goods (e.g., friendship or love), we find knowledge to be of *instrumental* and *final* value. We entertain VKI for both the cases where knowledge seems to be instrumentally valuable and for the cases where it seems to be finally valuable.

On the one hand, knowledge is found to be instrumentally valuable because it can function as a means to successful action, i.e., achieving our goals like satisfying desires, intentions, fulfilling plans and the like. For example, if I desire a glass of water then, trivially, to satisfy my desire is necessary that I have the relevant means-end knowledge that will guide me to satisfy my desire. If I am to satisfy this desire I must know where I can find a glass of water; otherwise, I rest my hopes for satisfying this desire on accidentally coming across water and this is unfortunate, as we surrender to the unpredictable hands of luck.

On the other hand, we also seem to find knowledge to be finally valuable. That is, valuable for its own sake and not as a means for something other than knowledge itself. For example, we value knowledge of how to solve a differential equation, a geometrical problem, prove a logical theorem etc. Such instances of knowledge are valued for their own sake and not for something practical like the guidance they can offer. For, obviously, such instances of knowledge cannot offer much of practical guidance in any immediate or direct sense. The fact that we seem to often engage in inquiry for the mere sake of cognitive achievement illustrates the point, that is, it illustrates that we often value knowledge for its own sake, regardless of practical considerations.[8]

More generally, we value practical knowledge but also value knowledge that is detached from the buzz of our practical lives. Whole branches of pure mathematics attest to that as they seem to have nothing direct to do with practicality and the same could be claimed about some branches of philosophy like abstract metaphysics (e.g., the debates on the nature of universals, the reality of time, etc.). Still, we consider such intellectual pursuits worthwhile and the derived knowledge valuable for its own sake, independently of having any direct practical purport.

On the basis of this analysis of VKI, we can distinguish between two senses of epistemic value: *practical value* and *pure epistemic value*. Practical value is instrumental value while pure epistemic value is final value. On the grounds that we entertain VKI in both senses, an adequate account of CVP should explain for both senses of epistemic value. Any theory that accounts only for one of the two senses is to be considered inadequate, as it would strike us as elliptical. It would leave out of the theoretical picture one of the two intuitive senses of epistemic value, and this would seem inadequate.

But it is worthy of notice that it would be inadequate only for a solution that attempts to vindicate and not explain away the intuitions built into CVP. Approaches that do not vindicate but explain away at least some of the intuitions built into CVP (and the rest epistemic value problems) would be *revisionary*. But "revisionary" here should be taken as neither pejorative nor honorific. A revisionary approach might be good or bad enough on independent

---

[8] Some philosophers have thought that knowledge is of final value because, exactly, it is a cognitive achievement. But there are problems for such a view as there are cases where, intuitively, we have knowledge but not a cognitive achievement and vice versa. See Pritchard, Millar and Haddoch (2010, ch. 2). for a criticism of this approach to final value.

grounds that have nothing to do with evolutionary considerations about the psychological mechanism operating behind VKI.[9]

At any event, what is to our interest here is not CVP but an evolutionary psychological explanation of the mechanism operating behind VKI. And if our analysis of the two senses in which we entertain VKI is correct, then an adequate psychological explanation must explain both the intuition that knowledge is valued instrumentally for its practicality and finally for its own sake, regardless of any practical value.

An evolutionary psychological explanation that does not explain why we are disposed to find knowledge valuable in both senses would leave something essential out and would therefore fail as an explanation. It would fail to inform our understanding of the psychological mechanism operating behind VKI and giving rise to CVP (and the rest of the epistemic value problems). For these reasons, I take this requirement to be a *desideratum* for any adequate psychological explanation of the mechanism that operates behind VKI.

One last important point remains before we embark on our evolutionary exploration of VKI. We should make no mistake about the theoretical scope of exploring an evolutionary psychological explanation of VKI. The *psychological* question of why we pervasively have VKI is quite independent of the *normative* question of what the value of knowledge really is. The question I aspire to tentatively explore is why knowledge appears valuable to us and not why knowledge really is valuable. We should be careful enough to discern that even if the evolutionary ruminations on VKI are to the right direction, this says nothing direct about "the core value problem" (and the other epistemic value problems).

We should make no mistake to claim the opposite, namely, that evolutionary considerations could on their own explain the value of knowledge because that would mean to breach Hume's (1985) famous "is/ought gap" and commit Moore's (2000) "naturalistic fallacy." This is what we may call "the Moorean/Humean lesson." Let me very briefly explain what "the Moorean/Humean lesson" is about.[10]

---

[9] Usually, what revisionary approaches to epistemic value try to do is to claim that instrumental value exhausts the value of knowledge and, thereby, explains away the final value of knowledge. This does not deny the reality of the intuition of final value, of course; what it denies is that we should take the final value intuition at face value. For such an approach see Ridge's PP presentation "Getting Lost on the Road to Larissa." An evolutionary psychological explanation of the value of knowledge intuition could lend a hand to revisionist approaches to epistemic value (like Ridge's epistemic value minimalism) because this would allow such people to explain why we have the value of knowledge intuition in the final value sense, though, this intuition should not be taken at face value.

[10] A similar point is found in Lewens (2007, 159-162). Also, philosophers who have applied evolutionary considerations on normative domains like knowledge and meaning, such as Craig (1990, 9) and Papineau (2003, 11), are careful enough to make clear that these evolutionary

As Hume (1985) has famously argued, you cannot deduce a normative injunction from merely descriptive facts like, among others, our evolutionary history as natural animals. You cannot deduce an "ought" from an "is," as philosophers sometimes say. No doubt, Hume didn't (and couldn't) have in mind evolutionary theory, as Darwin came almost a century after him, but his lesson still applies.

For example, the fact that wishful thinking, self-deception and other forms of doxastic irrationality may have been evolutionary advantageous for broadly pragmatic reasons does not mean that we ought to believe what is produced by the cognitive processes of wishful thinking, self-deception etc. Equally, the fact that wrongdoing like lying, stealing, etc. may have been evolutionary advantageous for broadly pragmatic reasons does not mean that we ought to lie, steal etc.

In a parallel with Hume, Moore (2000) with his "open question argument" argued that attempts to reduce moral properties to descriptive (or natural) properties commit "the naturalistic fallacy."[11] He argued that attempted reductions of moral properties to natural properties always meet "open feel" semantic intuitions that undermine such attempts. Moore himself applied "the open question argument" to the Social Darwinism of Herbert Spencer with a devastating effect. He argued cogently that evolutionary considerations couldn't reduce goodness. As he put it:

> The survival of the fittest does *not* mean, as one might suppose, the survival of what is fittest to fulfil a good purpose—best adapted to a good end: at the last, it means merely the survival of the fittest to survive; and the value of the scientific theory [of evolution], and it is a theory of great value, just consists in shewing what are the causes which produce certain biological effects. Whether these effects are good or bad, it cannot pretend to judge. (Moore 2000, 99; emphasis in the original)[12]

In essence, what "the Moorean/Humean lesson" teaches is that you cannot deduce what you ought to believe or what you ought to value from mere

---

considerations do not answer the normative questions of how we ought to use the concept of knowledge or what we ought to mean. As Papineau says "As a teleosemanticist I hold that our beliefs have been biologically designed to track their truth conditions. But I don't think that this does anything to show that they *ought* to do this" (Papineau 2003, 11; emphasis in the original).

[11] As Frankena (1939) argued, "the naturalistic fallacy" is not a *logical* fallacy. There is nothing incoherent in the idea that goodness can be reduced but we haven't yet found the right analysis. But philosophers who accept Moore's "open question argument" treat it not as a conclusive argument but as an inference to the best explanation for our "open feel" semantic intuitions that undermine such reductive efforts.

[12] Not all philosophers accept the Moorean/Humean lesson as there are philosophers who attempt to show that we can reduce moral concepts and bridge the gap between "is" and "ought." One example is Smith (1994). Here I will simply assume that such attempts are not successful and treat "the Moorean/Humean lesson" as a fact.

descriptive facts like our evolutionary history as natural animals. For this reason, even if there is a good evolutionary psychological explanation of why we have the pervasive VKI this does not answer why knowledge is valuable at all. It might overall *contribute* to such an answer but this will have to be part of a broader philosophical theory.

More generally, the moral of "the Moorean/Humean lesson" is that even if evolutionary considerations could inform our understanding of the origins and functions of our psychological and cognitive capacities, it won't be sufficient on its own grounds to answer normative philosophical questions of the familiar sort: What we ought to value? How we ought to live? How we ought to reason? What we ought to do? What we ought to believe? etc. Valuable the evolutionary theory maybe, it has certain theoretical limits that one should be cautious enough not to transgress.

The theoretical scope of such an evolutionary psychological explanation clarified, our evolutionary psychological ruminations in the next section 3 will attempt to explain both senses in which we find knowledge valuable and meet the imposed desideratum.

3 Evolutionary Ruminations on VKI

Evolutionary psychology as such is a relatively recently founded branch of psychology, although its roots go back to Darwin's work.[13] Roughly, it attempts to illuminate the operation of psychological and cognitive mechanisms by appeals to evolutionary considerations. That is, considerations invoking the idea that these mechanisms might have been genetically inherited due to the shaping hand of biological adaptation through natural selection. With this project in mind, evolutionary psychologists often resemble the mind with a Swiss army knife, namely, a knife containing multiple tools that can serve different purposes. Each tool's function has been naturally selected under adaptation pressures to play an evolutionary advantageous role.

But like the rest of evolutionary theory, evolutionary psychology's scientific status is controversial as some philosophers and psychologists tend to think that evolutionary psychological explanations are nothing more than "just so stories."[14] That is, they are theoretical stories that in reality remain highly speculative assertions with not much of substantial evidential support.

As my chief purpose here is to argue *from* the vantage point of evolutionary psychology rather than argue *for* evolutionary psychology, in what follows I will

---

[13] See Lewens (2007, chap. 5) and Buller (2007) for a discussion of the origins of evolutionary psychology.

[14] See Lewens (2007, 128-129) for a brief discussion of challenges to evolutionary psychology.

assume that evolutionary psychology could potentially inform our understanding of the origins and function of our psychological and other cognitive mechanism and processes. More generally, I will assume that it could inform our understanding of the functional structure of our cognitive architecture. No doubt, this is an assumption that needs to be argued for but arguing for this assumption here would have taken us far beyond from what the scope of this essay allows.

Let us now apply evolutionary considerations on the psychological mechanism operating behind VKI. As we made clear in section 2, evolutionary considerations should be capable of explaining why we are disposed to value knowledge for both its practical value sense and for its purely epistemic value sense. This is what we identified as a desideratum for any adequate psychological explanation of the mechanism operating behind VKI.

Now, the intuition I want to tentatively press is that the psychological mechanism disposing us to find knowledge valuable is something that has evolved to be a constitutive feature of our psychological architecture. It has evolved because it has been evolutionary advantageous for our struggling-to-survive hunter-gatherer ancestors of the Pleistocene period—the era spanning 1.8 million to 10,000 years ago, which is taken to be the formative period for evolving adaptations.[15] The idea is that we, the members of *Homo sapiens*, find knowledge valuable so naturally and effortlessly, because during our evolutionary history our hunters-gatherers ancestors who did entertain VKI were in better survival terms than those who didn't.

According to this idea, the psychological mechanism behind VKI has evolved to be a constitutive feature of our psychological architecture because it was chosen by means of the Darwinian natural selection due to adaptation pressures. Our ancestors that did have this psychological mechanism could better adapt and cope with the challenges of their natural environment while our ancestors that didn't have this psychological mechanism fared significantly worst in terms of adaptation to their natural environment.

Agents equipped with this psychological mechanism could adapt and fare better than agents that weren't equipped with this psychological mechanism for a very simple and intuitive reason. The reason is that being disposed to find knowledge valuable would naturally have been coupled with a desire for knowledge for its own sake. We usually have a desire for things we consider good and

---

[15] It is taken to be the formative period for evolving adaptations because our ancestors spent only the past 10,000 years living as agriculturists and the past few hundred years living in industrial societies. Given that the last 10,000 years our ancestors didn't meet much of evolutionary challenges as agriculturists, it is rather improbable that humans have evolved adaptations to post-pleistocene environments. See Buller (2008, 259-260) and Charlesworth and Charlesworth (2003, chap. 1) for discussion.

if we found knowledge to be good then it is reasonable to surmise that we had a desire for knowledge for its own sake.[16]

Thus, our ancestors that had this VKI and the desire for knowledge for its own sake, reasonably, would have channelled this desire towards practical knowledge concerned with how to satisfy their pressing sustenance needs for food, drink, shelter, clothing etc. and this would have allowed them to amplify their survival chances. I say they would have "reasonably" channelled this desire for knowledge towards practical knowledge of how to satisfy basic sustenance needs because they would also have the instinctive desire for self-preservation. They would have been disposed to desire to survive and to achieve this they would have to use practical knowledge for the satisfaction of their basic sustenance needs.

An evolutionary psychological explanation could again be given for our pervasive and almost involuntary entertainment of the instinctive desire of self-preservation. It is not difficult to imagine how this evolutionary explanation would go. Our ancestors that had this instinctive desire for self-preservation would have had better chances of survival from those that didn't because they would have taken more interest in themselves and their survival. Agents that didn't have the instinctive desire for self-preservation would have had substantially less chances of survival as they would have taken less or even no interest in themselves and their survival.

This understanding of the desire for self-preservation, though, should not give the wrong signals. It should not be assumed that our ancestors were pretty much Hobbesian egoists thinking only of themselves. For, there is nothing inconsistent in having both the instinct of self-preservation and other-regarding instincts (altruistic instincts, sympathy etc.). Actually, for reasons we need not pursue here it is quite plausible that there may be an evolutionary explanation for the reality of such other-regarding instincts as interpersonal cooperation and reciprocity would have often proved to be mutually beneficent for the agents of a community.[17] An agent, therefore, may very well have both, even though these may psychologically conflict on certain occasions.

If, then, our ancestors had this desire for knowledge for its own sake and channelled to some substantial extent this desire towards practical knowledge of how to satisfy their pressing sustenance needs (due to the desire for self-preservation), then they would have significantly enhanced their chances of

---

[16] There can be an evolutionary explanation here for why we tend to desire what we take to be good. If we act on the principle of what we take to be good and action requires a desire, as the prominent Humean theory of action suggests, then if we didn't tend to desire what we take to be good we would be rendered practically paralysed and this, clearly, would mitigate our chances of survival. For an influential defence of the Humean theory of action see Smith (1994).

[17] See Ernst (2008) for how evolutionary game theory approaches the phenomenon of altruism.

survival. Obviously, if you are motivated to amass knowledge of where abundant prey is, trees with edible fruit, what sort of mushrooms are poisonous, how to make traps, use a bow, skin a bear to use its fur, etc., you amplify your chances for survival and reproduction because it is more likely that you will succeed to satisfy your basic needs.

Instead, our ancestors that weren't equipped with such a psychological mechanism and didn't have such an instilled disposition to value and desire knowledge for its own sake would have significantly less chances of survival in the hostile environment of the war of nature; even if they did have the desire for self-preservation. Reasonably, if you are not much motivated to amass knowledge about where you can find clear water, fresh fruit, possible places like caves that would function as sheltering positions or hideouts in a case of emergency, which areas host dangerous predators like lions, how you can hunt, make traps, use tools etc. then your chances for survival are much less than one who thirsts for knowledge.

This evolutionary explanation seems to explain why we find various instances of knowledge to be finally valuable, that is, valuable for their own sakes without any direct practical purport. We have the intuition that knowledge is valuable for its own sake because we acquire by means of genetic inheritance the intuition and desire for knowledge *per se*, that is, for its own sake.[18] We find various instances of knowledge valuable in the final value sense because we have a psychological mechanism disposing us to value knowledge for its own sake and this mechanism is constitutive of our cognitive architecture for evolutionary reasons. It has been ingrained in our architecture by natural selection because of adaptive pressures.

If this is to the right direction, then we have a nice evolutionary explanation why VKI is often entertained in the final value sense. Still, this says nothing of why the value of knowledge intuition is often entertained in the practical value sense and, as we have diagnosed in section 2, any adequate explanation of the psychological mechanism operating behind VKI should explain why we entertain VKI in both senses. Otherwise, it would be incomplete and thereby inadequate.

Perhaps, the simplest and the most elegant way to explain why we have the intuition that knowledge is also practically valuable is not to invoke any further evolutionary considerations, but instead to invoke learning processes, as these are studied in cognitive and developmental psychology. That is, we should consider how our instinctive disposition to value and desire knowledge *interacts* with learning processes to provide us with the practical value sense of VKI. If this idea goes in the right direction, then our intuition that instances

---

[18] Papineau (2003, 73-80) makes a parallel point about how evolution could have selected a desire for true belief *per se*, as he says.

of knowledge are valuable in the practical sense is the product of interaction between our instinctive disposition to find knowledge valuable for its own sake and learning processes. But let us belabour the idea first.

Let's start from what seem to be empirical facts. It seems to be a fact that from infancy we find ourselves oriented towards satisfying our basic sustenance needs. We cry if we are not fed or watered on time, if we are cold etc. One idea then is that as animals with biological needs from tender infancy we exhibit the disposition to channel our desire for knowledge towards knowledge of how to satisfy our sustenance desires. We exhibit such disposition because, as we have canvassed, we also seem to have the instinctive disposition for self-preservation. We instinctively cling on what is life-preserving and refrain from what is life-endangering. We indulge, for example, in pleasure and refrain from despicable pain.

Such practical knowledge is surely to be considered very valuable as it is substantial for survival. For example, if as infants our crying is efficient in making our parents satisfy our desires then, in some sense, we learn by association that with crying we can satisfy our desires and we come to value crying and deploy it when we see fit. This is why infants seem to be particularly spoilt as they cry when their desires are not readily satisfied.

But this happens during infancy. As we grow up and acquire a first natural language and conceptual powers in the context of a community, we go on seeing practical knowledge as something good because it allows us to satisfy desires that often go far beyond the basic sustenance desires. Among other things, we desire a TV, a computer, a car and many other technological products that advertisements bombard us with in the settings of our modern society.

Thus, in time we come to associate practical knowledge with value and acquire the unconscious habit to consider knowledge that allows us to satisfy our desires as something good. We come to habitually correlate practical knowledge with value and see practical knowledge as something valuable. Actually, the habit is so deeply internalized and unconscious that it takes some philosophical reflection to acknowledge its underlying reality. We need to exercise our conceptual powers for reflection in order to acknowledge the reality of the unconscious habit of association between practical knowledge and value.

If this psychological explanation is to the right direction, then explaining the psychological mechanism operating behind VKI is partly evolutionary and partly cognitive and developmental. It is evolutionary to the extent we have an innate disposition to value and desire knowledge per se and is both evolutionary and cognitive-developmental to the extent we channel this desire towards practical forms of knowledge because of the self-preservation instinct. The product of this interaction is the formation of a deeply rooted, unconscious habit to see knowledge as practically valuable because it allows us to satisfy our desires, fulfil our plans etc.

In conclusion, Aristotle's famous opening sentence of his *Metaphysics*, namely, that "all men by nature desire to know" is quite to the point, albeit, for evolutionary reasons that in all evidence Aristotle, despite his teleological understanding of nature, was not aware of. For, although Aristotle was often saying that "nature does nothing in vain" implying that things have a *telos*, a functional purpose they are made to serve he was still unaware that at least many of these functional purposes have been shaped by the mechanism of natural selection. A mechanism whose operation Darwin first famously made explicit.

4 A Review of the Argument and Conclusion

The goal of this essay was to engage on exploratory evolutionary ruminations on the origins and function of the psychological mechanism operating behind VKI, that is, our disposition of finding knowledge valuable. Such an evolutionary psychological explanation would account why we human beings find so pervasively knowledge to be valuable.

In the introductory section 1, I made explicit my goal and outlined the structure and content of the essay. In section 2, I explained how VKI gives rise to CVP, and in what senses knowledge is to be considered valuable, namely, instrumentally and finally. Afterwards, I explained than any adequate psychological explanation of VKI should account for both senses in which we entertain VKI and took this to be a desideratum for any adequate psychological explanation of VKI.

In section 3, I applied evolutionary consideration on the psychological mechanism operating behind VKI. By appeal to evolutionary considerations I attempted to account both for the practical sense and the pure epistemic sense of the value of knowledge, as the desideratum identified in section 3 prescribed.

As I explained, we could speculate that the psychological mechanism behind our intuition to value knowledge both practically and for its own sake has evolved to be a constitutive feature of our psychological architecture because it was chosen by means of Darwinian natural selection due to adaptation pressures. We have been endowed, courtesy of Mother Nature, with a psychological mechanism that disposes us to find knowledge valuable for its own sake because such a psychological mechanism would amplify our ancestor's chances for survival and reproduction.

It would have amplified our ancestors' chances for survival and reproduction because such disposition to value knowledge would have been wed with a desire for knowledge and this coupled with the instinctive desire for self-preservation would have channelled our desire for knowledge per se towards practical knowledge of how to satisfy our basic biological needs.

This evolutionary psychological explanation seems to explain the origins and function of the psychological mechanism operating behind the disposition to find knowledge valuable. Still, this explains only why we find knowledge valuable for its own sake as the psychological mechanism disposes us to find knowledge valuable for its own sake. To account for the practical value sense of VKI, I have appealed to the interaction between the disposition to find knowledge valuable for its own sake and learning processes as these are being studied by cognitive and developmental psychology.

The idea was that we come to have the practical value sense of VKI because from infancy, due to the self-preservation instinct, we learn that knowledge of how to satisfy our basic biological needs is valuable. As we grow up and our desires multiply and go beyond our basic biological needs, we come to form a deeply rooted, unconscious habit to associate practical knowledge with value because such knowledge allows us to satisfy our desires.

References

Aristotle (2003). *Metaphysics Books I-IX*. Trans. by H. Tredenick. Harvard University Press.
Buller, D. (2007). Varieties of Evolutionary Psychology. In Hull and Ruse (2007), 255-274.
Charlesworth, B. and Charlesworth, D. (2003). *Evolution*. Oxford: Oxford University Press.
Craig, E. (1990). *Knowledge and the State of Nature*. Oxford: Oxford University Press.
Ernst, Z. (2007). Game Theory in Evolutionary Biology. In Hull and Ruse (2007), 304-323.
Fodor, J. (1983). *The Modularity of Mind*. Cambridge, Mass.: MIT Press.
Frankena, W. (1939). The Naturalistic Fallacy. *Mind* 48.
Hull, D. L. and Ruse, M. (eds.) (2007). *The Cambridge Companion to the Philosophy of Biology*. Cambridge: Cambridge University Press.
Hume, D. (1985). *A Treatise of Human Nature*. Harmondsworth: Penguin.
Kvanvig, J. (2003). *The Value of Knowledge and the Pursuit of Understanding*. Cambridge: Cambridge University Press.
Lewens, T. (2007). *Darwin*. London: Routledge.
Moore, G. E. (2000). *Principia Ethica*. Cambridge: Cambridge University Press.
Papinaeu, D. (2003). *The Roots of Reason*. Oxford: Oxford University Press.
Plato (1999). *Meno*. Harvard University Press. Translated by W.R.M.Lamb.
_____ (2006). *Philebus*. Trans. by H. N. Fowler and W. R. M. Lamb. Cambridge, Mas.: Harvard University Press.
Pritchard, D. (2007). The Value of Knowledge. In *Stanford Encyclopedia of Philosophy*. E. Zalta (ed.). http://plato.stanford.edu/entries/knowledge-value/
_____, A. Millar and A. Haddock (2010). *The Nature and Value of Knowledge: Three Investigations*. Oxford: Oxford University Press.
Ridge, M. (2009). Getting Lost on the Road to Larissa. PP presentation at the University of Edinburgh.

Smith, M. (1994). *The Moral Problem*. Oxford: Blackwell.
Tooby, J. and Cosmides, L. (1992). The Psychological Foundations of Culture. In *The Adaptive Mind*. J. H. Barkow, J. Tooby and L. Cosmides (eds.). Oxford: Oxford University Press.
Williamson, T. (2000). *Knowledge and its Limits.* Oxford: Oxford University Press.
Zagzebski, L. (1996). *Virtues of the Mind.* Cambridge: Cambridge University Press.

12
Teleology as a Theory of Meaning
Juraj Hvorecký

1 Introduction

Understanding the significance of the teleological project requires viewing it as part of a larger project of the naturalisation of semantics. This larger project aspires to an uneasy task of explicating intentional terms, such as representation, being about or standing for something else, via naturalistic vocabulary. Difficulties, immanent in the project of naturalisation of intentionality, are well known. Intentional relations apparently transcend the obvious limitations of space and time. We can think of past events, contemplate a distant future and speak of fictional characters. In all these scenarios, our mental processes are intentionally related to some entities, their intentional objects. Yet a mode of existence of these objects is very unclear and at first sight unlikely to be explicated by any naturalistic framework. Conversely, naturalism employs terms borrowed from sciences or generalizes from them and at the same time it is quite obvious that science rejects interactions with past, future and possible events. Natural phenomena are taking place within strict temporal and spatial boundaries, governed by laws of nature. So while intentional relations are free-floating, transcending all limitations and omnipresent in mental episodes, the natural world is bound by principles that are law-like and strict. This tension calls for an explication. With naturalism aspiring to explicate mental domain as a fully credible category, it has to attempt to subsume intentional terms under a natural order and "bring the phenomenon of representation within the scope of the natural sciences" (Papineau 2006, 175).

Any theory of intentionality, whether naturalistic or not, takes as a basic starting point the assumption that two elements comprise all representational relations: what is represented and what represents. What represents takes in what is representing and presents it as its object. In our case of mental states, thoughts, dreams and contemplations are playing the representing part, while what is represented are particular contents that mental states incorporate. A dream is about a lottery win, representing the win as something the dream by being about it. Such a dual structure constitutes the cornerstone of intentionality and all aspirants that want to say something substantial about the subject matter have to explicate these foundations.

The talk of teleology as a part of a larger project of representational naturalism implies there are other competing naturalistic views on intentional phenomena. Because our central topic deals solely with challenges to teleology, we will just briefly enumerate its competitors. Various versions of causal theories, inferential role semantics and success semantics are all in the ballpark (for an overview, see Papineau 2006). In their variety, all these theories share naturalistic intuitions, but differ in their perspectives on which naturalistic terms to employ in an explanation of intentional facts. There is one additional feature of all competing theories that is worth stressing. It is an assumption that the basic bearers of intentionality are mental states. While other states or entities can also represent and stand for other things, they do so in a manner that depends on mental states having such capabilities. All non-mental intentionality depends on its mental adumbration. A picture can represent a feature of the world, but it does so because its representational powers are derived from mental states of both its creator and the observing public. However, the shared intuition over the mental origin of intentionality does not prevent advocates of various competing theories to disagree strongly on basic facets of the naturalistic project. Differences in opinion over the precise nature of reducing intentionality have led to heated debates, though they have been also marked by significant progress in this domain. In fact, Papineau has acknowledged that teleological theories have arisen from the dissatisfaction with the achievements of competition.

2 Biological Basics of Teleology

Before moving to the contentious issues, let us first turn into an exposition of how exactly teleology attempts to naturalize intentionality. In explicating its dual structure, we will begin with what is represented, i.e. the content of the intentional state. This is a common strategy in most versions of naturalized semantics. In non-teleological versions it is assumed that uncovering how the object of representation is linked to a representation bearer is sufficient to explain the very essence of intentionality. Once we discover how the represented side is tied with the representing one, and in doing so employ only broadly scientific terms, the puzzle of temporally and spatially unconstrained relation disappears. Explaining this link suffices for the explication of intentional relation.

Teleological theories utilize a different approach. Instead of stressing the link between content and a corresponding representative entity, teleologists elucidate the content via a function of the state that carries it. While its competitors seek for a naturalistic link between a state and its object, in teleology, content is determined by a link that is external to the very core of the representational relation. A *function* of the representing state, i.e. what the state does, is essential for specifying its content. Yet there are many functions any state executes, and quite a few of these can be explained within naturalistic ontology. Teleologists need to concentrate on a specific subset of state's functions. This subset that is to shed a light on content picks up *biological* functions of states as unique content determinants. An emphasis on biological functions of the state constitutes the most significant demarcation between teleological and all other contending theories in the field of naturalized semantics. Unlike its competitors, teleology ties content determinants with functional biological underpinnings of states that carry corresponding contents. A novel link is added to the one between content and its bearer. In it, functional attributes of the latter are employed to elucidate the specification of the former. The consecutive step is then to show how representational content of mental states can be explained in terms of their biological functions.

Before making the next step by moving onto the explicatory core of the theory, the concept of a biological function has to be clarified. Biological functions are properties of biological traits. It follows from the theory of natural selection that if a biological trait has a certain function, this is because the function produces an advantageous effect to the organism. Function occurrence has emerged and is maintained solely because of these advantageous effects. As Papineau puts it: "biological functions are in the first instance always a matter of effects" (Papineau 2006, 181). Had there been no matching effects, the function would either not have emerged or have ceased to exist. In what follows, we will neither be concerned with what the advantage consists of, nor how to measure the advantage. Instead, we will concentrate on the notion of function effects and what they bring along.

Given that teleological explanations intend to explicate content in terms of biological functions and these, in turn, are a matter of particular effects, it is not surprising that effects of content bearers are crucial for determining their content. This observation is central to understanding the novelty of teleological strategy. It was Ruth Millikan who first appreciated consequences of this specific point. In her seminal 1989 paper Millikan argues that once we take the notion of biological functions in explaining representative relations seriously, we can no longer sustain the dual intentional structure. Duality of what is represented and what represents prevents us from understanding crucial aspects of representational relations. Suggested insufficiency becomes visible when one considers the fact that the dual structure has no place for any effects of the representing state. When the duality is taken for granted, intentionality is assumed to be only arising from whatever constituted the relation between bearers of representations and their content. No further element has to be taken into account.

Instead of building a theory around two components of intentionality, Millikan calls for an introduction of a third component that maps the effects of representing. The added component has an advantage to shed a light on two previously known ones. With biological functions individuated by their effects, the third component has to reflect the effective side of representations. It thereby comes "after" the dual structure, but its position should not be understood in temporal or spatial terms. Rather, its position is best understood in functional and causal terms.

In her introduction of the third element, Millikan opts for what we might term the "capitalist terminology." Only one item in the new triad retains its name from the original dual structure model and this fact is indicative of its rather loose place in the new schema. According to the new terminology, the represented is accompanied by a producer and a consumer of representations. More important than terms are the roles played by the representation producers and consumers: "It is the devices that use representations which determine these to be representations and, at the same time, determine their content" (1989, 291). The quote indicates that consumption of representations determines representational powers of an entity and also what precisely it stands for. The other way to think of the consumption is to restate its determinants as truth conditions. The representation is true in virtue of how it promulgates its content. As Papineau puts it: "we can think of the representation's truth condition as the circumstance that enables it to fulfil [its] function" (2006, 177). Instead of content determination taking place by the representing parts to stand for the represented in an appropriate relation, the representing state offers its content for a further use. It is the mechanism of a further consumption that specifies that content has been received and its specific characteristics detected: "a good look at the consumer part of the system ought to be all that is needed to determine not only representational status, but also representational content" (Millikan 1989, 293). Although the quote might give an appearance of the consumer component being primarily an epistemic device, the point is far from being purely epistemological. Without entities that consume representation there simply would not be any representational relations at all.

How can a consuming device determine a content of something external to it? This question is a bit misleading. Emphasizing the external position of a consumer device is tantamount to advocating the old dual model. For teleologists, the consumer device is an essential part of an act of representing and thereby it is internal to the process of intentionality. It doesn't stand alongside the old dual structure, but rather makes the dual understanding possible. However, despite the rejection of the externality, it is true that we need to learn more about ways in which the third component operates. The consumer device determines the content by its reception and a subsequent spectrum of actions that the device is prompted to execute upon its reception. It is the after-effects of what happens once representation is consumed that really matters. The form of what is received from the sender determines what was sent. Determination of the content depends on a variety of outcomes that follow the intake and manipulations brought about by the consumer device. An example might illustrate this point.

On a standard dual structure account, a phone ring represents an incoming call because there is an intrinsic link between the two, established by the designer of the phone. Whenever there is a call, there is a ring. The established link between the call and the ring is enough to mark their relation as indicative. According to teleological accounts, something is missing in this link to count as fully intentional. The missing part is the effect caused by the ring. It is because what one does after hearing a phone ringing that the ring stands for an incoming call. An available spectrum of actions means that the ring stands for a phone call. Had there never been a reaction from anyone after the prolonged and repeated beeps, the representational state would not obtain. The lack of meaning would persist even if there was some original designer that intended phone rings to indicate calls. The absence of intentionality is definitive because of the lack of effects. Without any consumption the carried content cannot come forward. The third structural element is crucial for the relation to obtain, as "teleological strategy requires us first to identify which results the state is supposed to produce, and then use this to tell us what it is representing" (Papineau 1998, 10).

The example with the phone ring should not be taken as a paradigm case for all scenarios. Its ultimate drawback is a definite presence of a designer. There is someone who imposed a function of calling and drawing attention to an incoming call onto the apparatus. Defenders of the dual model of intentionality could claim that the meaning of the ring was bestowed on it by its very design. It stands for the incoming call, regardless of occurrences of any after-effects. The connection between ringing and the call is a matter of design rather than any consequence that the event might bring about.

While such an objection can be valid in the domain of artefacts, no similar helping hand is available in the natural world. Teleology presupposes evolutionary facts and is primarily concerned with explaining occurrences of representations in the living systems where the premise of a designed meaning is ruled out. Advocacy of any supreme designer would compromise the naturalistic outlook of the entire programme.

3 Indeterminacy of Selected Traits

An introduction of the consumer element and a move to emphasise effects in explaining representational relations clearly amount to one of the most inspiring moves in the naturalization debate. Yet the argument to shift the focus on consequences of representations didn't go unchallenged. A powerful attack was waged by Jerry Fodor (1990, ch. 3) and various counter-arguments quickly followed. A brief summary of his line of argument goes like this. Given that the notion of biological function is defined via effects, there should be a co-variation between the carried content and the conditions, upon which the function, assigning the content, is fulfilled. Yet many biological functions are fulfilled by achieving goals that cannot in retrospect adjudicate an ascription of one particular content over another. This might seems as a rather innocent observation, but the consequences for the programme of naturalized semantics are plain disastrous. Representations carried by functional states with ambiguous consequences remain indeterminate. The consumer device cannot specify the content to a desirable degree. Narrowing down contents of representations turns out to be an impossible task. On a standard account of representation this result is very unwelcoming and it never occurs on dual models where a causal link to whatever is represented is always determinate. Familiar examples of representational contents are always fully specified. My belief that the summer is hot takes as its object a proposition that *the summer is hot* and no other. In many cases, intentional objects are more precisely specified than the real world entities they are related to (as demonstrated, among others, by Frege in his discussion of modes of presentation).

Fodor gives an example that everyone in the field feels obliged to comment upon. Supposedly, frogs stick their tongues out in order to catch flies and they use simple detection mechanism for timing their muscular movement. It is therefore assumed that within their cognitive system a representation is carried from the visual brain centres to tongue-sticking centres. This presumed representation carries certain content. Unfortunately, it is by no means obvious that the content of this cognitive state, associated with catching flies, does in fact operate over the content *fly*. The reason for scepticism lies in the fact that frogs stick their tongues out whenever any small black object crosses their visual field. What is carried to the sticking-out brain centre has to be specified in a more liberal manner, otherwise the tongue would never shoot at non-flies. As a matter of fact, most small black flying objects in the frog's *Umwelt* turn out to be flies, yet whether the frog conceives of them under that particular description is quite doubtful. Concentrating on the effects of fly-catching offers little help. Flies are consumed as food, serve as a supply of protein and enhance reproductive chances of their catchers. All these functions can be carried by the initial cognitive state as its content. In fact, there seems to be no principled way to decide whether frog's content is about flies, food, small black flying objects or some other co-varying entity. The content of frog's internal states remains indeterminate and teleology, which is built as a theory of content, apparently fails in its primary task. It isn't able to deliver any elucidation on one of the basic elements of intentional relations. If teleology cannot succeed in specifying the intentional content, its destiny is in ruins.

Teleologists admit that Fodor's attack is a powerful one. However, they are not convinced that it should force them to abandon this philosophical project in its entirety. Instead, they come forward with a variety of defences of the teleological strategies. Agar (1993), Millikan (1991) and Neander (1995) present some of the early defensive responses, but we will concentrate on a strategy adopted by Papineau in (1998) and (2003).

4 Toward Determination

Papineau opens his response to Fodor with a close examination of both Neander's and Millikan's theories. While he disagrees with their respective conclusions, he takes inspiration from both of these approaches. Papineau's first charge against Fodor is his observation that not enough attention is paid to the consequential side of the representation. Looking at what follows the consumption of a representation is the best guide to narrow down its content. Its causal antecedent is irrelevant – therefore the content of the representation in question certainly cannot be assumed to be *fly* simply in virtue of a stipulated causal link. A decisive argument for the content determination is to be dug up from the aftermath of its detection. Millikan suggests that looking at advantageous effects the representation brings should simplify our search for the determinate content. In her view, *food* is the description under which the content of the frog's state can be best approximated (1991). It is because the most obvious advantageous effect of fly snapping leads to this particular result and as such it also points out to what the content is. While appreciating the consumer side of representative relations, Papineau is right to point out that Millikan fails to deliver a decisive argument to disarm Fodor (see his 1998, 5). Cutting the causal link between an entity and its representation rules out some candidates for the content ascription. Still others are kept untouched by such considerations and the problem of indeterminacy just won't go away. Papineau repeats Fodor's arguments against Millikan that there are many advantageous effects that content can deliver and, therefore, many possible content ascriptions. As I have indicated above, a reception of any particular content and the organism's reaction to it has both immediate and long-term effects and all of them count as advantageous. Tongue-sticking leads to a digestion of a fly, it enriches the body with proteins (and many other elements), promotes organism's well-being and enhances its chances to reproduce. As a result, the content of the state in question can be standing for any of these effects. We might with an equal force ascribe it as *food*, *protein supply* or *reproduction enhancer*. On Millikan's account, there is no principled way of distinguishing among a variety of options and content disambiguation once again fails once again.

Papineau then turns his attention to a proposed solution, first suggested by Neander (1995). She suggests that teleology has capacities to fully disambiguate the content, but instead of representation functions one should turn one's attention to representation malfunctions. When the trait is fully operational, its consequences are wide-ranging and that prevents backward content disambiguation. However, an inquiry into a hypothetical failure of systems that depend on the content delivery might further specify an exact function of a provider, thus allowing the content to be determined more adequately. Naturally, a failure of a system might have wide-reaching consequences on other systems that are coupled with a content producer and the threat of multiplicity is looming again. A failure in the frog's visual system results in failures to capture insect, digest appropriate food and could lead to extinction. Fortunately, there is a way to move forward with the helping hand of counterfactual thinking. In order to determine the content of the visual system, one has to concentrate on the most proximal malfunction that the collapse of the visual system leads to. The most proximal failure is indicative of a hierarchy among multiplicity of consequences: "the lower descriptions describe the underlying mechanism, and the higher levels explain why doing that was adaptive and selected for" (Neander 1995, 116). Problems in digestive system or reproductive success could be brought about by failures of a variety of preceding mechanisms, while only the function to detect small black objects fails solely because of damage to the visual system in question. The carried content is therefore *a small black object*.

Papineau appreciates Neander's use of malfunctions in the process of content determination, but he points out that Neander is not quite faithful to the teleological schema. It is hard to grasp the idea that a visual system produces the content *small black objects* given that small black objects are what the system detects. Detecting and being about are roughly synonymous terms: If anything, the link between *small black objects* and small black objects is causal. Concentrating on a unique source of a malfunction leads us, rather predictably, to the production side of a representation, instead of its consumption side. The feedforward constituency is not compatible with the teleological program.

Therefore, Papineau comes to a bit pessimistic conclusion. It is not in the power of teleology to fully determine content. However, it is not teleology, but rather an inappropriate target that leads to this failure. An assumption that content is fully determined in animals without belief-desire psychology is unjustified. Extrapolation from human cases, where content is always determinate due to its function in desires and beliefs, to animals with no equivalent functional roles, lies behind the overall failure of naturalism to determine content.

I do not accept Papineau's scepticism. On the one hand, it is open to further worries about the indeterminacy of content in humans that are suggested by Dennett (see Dennett 1987). Where Papineau points out an insufficient specificity of functions of traits in animal cases, Dennett suggests that the human belief-desire psychology might face similar problems, as it regularly calls for further communicative content specification. Therefore, content cannot be fully determinate at the very outset. Yet this is not the line of argument that I wish to pursue here. Rather, I want to point out that proto-psychological traits can be selected for as modules. Outputs of modules could be conceived of as concepts. They subsume certain input features under a specific category, thereby at least serve the function of recognitional capacities that concepts are supposed to provide. If this is the case, then the solution to the problem of content determination depends on a further philosophical worry about concept constitution. It seems that Papineau is committed to a view that rejects concepts as purely recognitional capacities and wants to see them integrated into an inferential network of beliefs. Such a view is popular, yet it runs counter to naturalistic intuitions that would like to see full-fledged content being built up from more basic elements and not to appear only with belief-desire psychology. In any case, settling the issue of the nature of concepts is beyond the scope of this chapter. Yet it appears to me that a solution to the determination problem should be attempted independently from the problem of nature of concepts.

Given that Papineau's critique of Neander on a seemingly unrelated philosophical problem, his position seems unfair. If the problem of Neander's account consists of her not being faithful to the spirit of teleology and yet the same spirit forces Papineau to deny content to all creatures without belief-desire psychology, then teleology seems to run dry. Naturalism has to turn to other options and Neander's elegant biological account that might rest partly on causal approaches is such an option. The very admission of employment of causality in an account of content determination, when employed moderately in counterfactual historical explanation of selective pressures, should not scare us. After all, causation is a perfectly legitimate feature of the natural world and when implicated in a biologically based account, it deserves our appropriate attention.

References

Agar, N. (1993). What Do Frogs Really Believe? *Australasian Journal of Philosophy* 71 (1): 1-12.

Dennett, D. (1987) *The Intentional Stance*. Cambridge, Mass.: The MIT Press.

Fodor, J. (1990). *A Theory of Content and Other Essays*. Cambridge, Mass.: The MIT Press.

Millikan, R. (1989). Biosemantics. *Journal of Philosophy* 86 (6): 281-297.

_____ (1991) Speaking up for Darwin. In *Meaning in Mind: Fodor and His Critics*. B. Lowever and G. Rey (eds.). Oxford: Blackwell, 151-164.

Neander, K. (1995). Misrepresenting and Malfunctioning. *Philosophical Studies* 79 (2): 109-141.

Papineau, D. (1998). Teleology and Indeterminism. *Australasian Journal of Philosophy* 76 (1): 1-14.

_____ (2003) Is Representation Rife? *Ratio* 16 (2): 107-123.

_____ (2006) Naturalist Theories of Meaning. In *Oxford Handbook of the Philosophy of Language*. E. Lepore and B. Smith (eds.). Oxford: Oxford University Press, 175-188.

13

Adaptationism, Deflationism and Anti-Individualism

Tomáš Hříbek

1 Introduction

A number of naturalistic philosophers have been trying to integrate theories of mental representation within the domain of evolution and biological function. In particular, these philosophers suggest that adaptationism, which is central to the Darwinian science, can provide a ground for anti-individualist, or externalist, accounts of representational states that have become dominant in philosophy of mind. Adaptationism is usually presented as an *empirical* claim about the causes of phenotypic traits. For starters, we can make do with Elliott Sober' definition:

> Natural selection has been the only important cause of most of the phenotypic traits found in most species. (Sober 1996, 72)

Sober is clear in treating adaptationism as an empirical thesis, albeit the one whose truth-value could be determined "only in the long run" (ibid.). However, some philosophers as well as scientists have meant something a lot stronger by "adaptationism." But we shall come back to that later. At any rate, it seems beyond doubt that psychological anti-individualism is definitely not a mere empirical claim. It is meant to be a *metaphysical* thesis about the *nature* of representational states. Most of traditional philosophy of mind, in virtue of its assumption that representational mental states can be fully characterized by attending solely to the properties and states internal to the individual's bearer of these states, has been individualistic. By contrast, anti-individualism is a relatively recent theory originated by Tyler Burge. According to his recent definition, anti-individualism is the claim that

> the natures of many mental states constitutively depend on relations between a subject matter beyond the individual and the individual that has the mental states, where relevant relations help determine specific natures of those states. (Burge 2010, 61)

Burge means that many mental states, specifically those that are representational, would not be what they are, unless the relevant relations to the external environment were in place. Therefore, the idea of a constitutive dependence of representational states of relations between the individual and her environment is distinct from the idea that mental states causally depend on such external relations. The latter idea is quite acceptable to many individualists. However, while agreeing that thoughts or perceptions are prompted by events in the external environment, individualists go on to claim that those mental states are then fully identifiable in terms of factors internal to the individual. Constitutive dependence is also stronger than metaphysical dependence. Necessities such as that it's true of each mental state that it is not a number or made of cheese are weaker than constitutive dependence because they do not figure in explanation of the nature of mental states.

These are the claims about the character of the relations between mental representations and their environment. The further issue to consider is whether the mind reduces to its external relations; in other words, whether it belongs in the ontological category of relation. Burge resists this conclusion with an example from biology: "It is constitutively necessary that to be a heart, an organ must have the function of pumping blood through a circulatory system" (ibid., 66). However,

> the relations to these other entities are not part of the internal structure of the heart. Nor is the heart itself a relation. Thus the nature of the heart is constitutively dependent for being what it is on relations to things beyond it. But the heart itself has a structure that is not made up of those relations. I think that representational mind is like that. (Ibid.)

Now, the idea behind the naturalization project under discussion here is to present anti-individualism as an implication of adaptationism, whether the latter is thoroughly empirical in nature or something else. Thus, Daniel Dennett claims that

> Burge's anti-individualistic thesis is then simply a special case of a very familiar observation: functional characterizations are relative not only to the embedding environment, but also to assumptions about optimality of design. (Dennett 1987, 310)

In other words, Dennett suggests that anti-individualism is implied by the fact that the ascription of representational states, which he construes as functional states in the biological sense, presupposes a description of the *selectionist history* of these states. It is precisely this appeal to history that is supposed to supply the required external component—i.e., something outside the individual—of the identity conditions of representational states. Likewise, in her elaborate theory of psychological explanation as a species of biofunctional explanation, Ruth Millikan argues that the functions in question are to be characterized historically. She says:

> I would like to explore implications for the science of psychology of the thesis that the categories of intentional psychology are function categories in the biologist's sense of "function," taking this to be a sense in which function is determined by evolutionary history rather than by current dispositions. (Millikan 1993, 171)

Thus, Dennett and Millikan share a conviction that anti-individualism can be vindicated as a feature of a psychology understood as an offshoot of the adaptationist program in biology. In short, their defence of anti-individualism assumes that psychological explanation is *of a kind* with biological explanation. In addition to the reductionist attitude towards psychology, contemporary naturalists also miss the metaphysical character of the thesis of anti-individualism. With an approving reference to Millikan's work on biological function, Dennett says that it is a particular individual's historical origin that "licenses a certain way of speaking" (Dennett 1987, 292) of her states. It is on the basis of the selectionist history of an organism or the design history of an artefact that we may describe certain of their states as serving the purpose of representing their environment. However, this means that Dennett and other naturalists interpret anti-individualism as a mere *semantical* thesis. That is, anti-individualism turns out to be a claim about how certain internal states of the individual ought to be described—namely, by including the mention of some facts external to the individual, in particular those about her history. For Dennett, this semantical reinterpretation of anti-individualism is part of his rejection of what he calls *original intentionality*. This is the claim that humans possess genuine minds while other candidates (e.g., computers and other artefacts) have minds only by proxy. Dennett argues that there are no minds at the bedrock. All intentionality is derived.

Burge rejects both the reduction of psychology to biology and of the metaphysical thesis to a semantical thesis. As for the former, in his recent book, *Origins of Objectivity*, he asserts:

> The explanatory content and goals of theories of perception and belief are not the same as those that underwrite biology. Explaining the way veridical and non-veridical representational states arise, given proximal stimulation, is a different explanatory enterprise from that of explaining any states in terms of their biological functions—their contributions to fitness. So biological explanations cannot reduce explanations whose point is to explain accuracy and inaccuracy of representational states. Since what they explain is different, the former cannot take over the job of the latter. (Burge 2010, 303)

As for the metaphysical character of anti-individualism, recall that Burge makes clear in his definition of the thesis that it concerns the very natures of mental representations. Thus, while no one can claim a monopoly on the usage of technical philosophical claims such as "anti-individualism," it appears that naturalists give it a substantially different meaning from Burge.

However, in this chapter, I do not wish so much as adjudicate the dispute between Burge and the Darwinian naturalists. Rather, I plan to analyze the arguments on both sides in order to understand the grounds of the disagreement better. I start with explicating the naturalists' strategy to derive anti-individualism from adaptationism (section 2). In the process, I note some important differences between Dennett and Millikan, especially over intentional realism. It turns out that Millikan does not share Dennett's view that all intentionality is merely derived; in this respect her position is closer to Burge's. And yet, Burge rejects Millikan's naturalism as well, on the grounds that her construal of representation is as far removed as Dennett's from the actual practice of psychology (section 4). However, I shall also point out a certain discrepancy in Burge's argument. We have already seen him appealing to a particular example of the biological organ, the heart, in explicating the character of the environmental dependence of mental representations. Elsewhere he used it to argue for the autonomous character of psychology. Now Burge's example is repeated almost verbatim by Millikan in her theory of biofunctional explanation. And, curiously, the example implies an individualistic understanding of psychology that Burge officially rejects (section 3).

2 Adaptationism and Anti-Individualism

Let us start by looking at the details of anti-individualist arguments. This should be interesting because both Burge and the naturalists appeal to very similar thought experiments. These are the notorious thought experiments featuring physically identical, yet intentionally distinct, individuals, whose intentional difference is explained in terms of a difference between the two individuals' social or physical environments. Burge summarizes both types of his thought experiment as follows:

> Consider a person A who thinks that aluminium is a light metal used in sailboat masts, and a person B who believes that he or she has arthritis in the thigh. We assume that A and B can pick out instances of aluminium and arthritis (respectively) and know many familiar general facts about aluminium and arthritis. A is, however, ignorant of aluminium's chemical structure and micro-properties. B is ignorant of the fact that arthritis cannot occur outside of joints. Now we can imagine counterfactual cases in which A and B's bodies have their same histories considered in isolation of their physical environments, but in which there are significant environmental differences from the actual situation. A's counterfactual environment lacks aluminium and has in its places a similar-looking light metal. B's counterfactual environment is such that no one has ever

isolated arthritis as a specific disease, or syndrome of diseases. In these cases, A would lack "aluminium thoughts" and B would lack "arthritis thoughts." Assuming natural developmental patterns, both would have different thoughts. (Burge 2007, 222-223)

Burge notes that the difference between the two individuals cannot be reduced to a difference in the causal origin of two tokens of the same type of a representation. Each token representation is of a different type because each has a different content. And the ascriptions of mental representations are "literal."

Dennett's story features artificial devices rather than human protagonists:

Consider a standard soft-drink vending machine, designed and built in the United States, and equipped with a transducer device for accepting and rejecting US quarters. Let's call such a device a two-bitser. Normally, when a quarter is inserted into a two-bitser, the two-bitser goes into a state, call it Q, which "means"(note the scare-quotes) "I perceive/accept a genuine US quarter now." Such two-bitsers are quite clever and sophisticated, but hardly foolproof. They do "make mistakes" (more scare-quotes). That is, unmetaphorically, sometimes they go into state Q when a slug or other foreign object is inserted in them, and sometimes they reject perfectly legal quarters—they fail to go into state Q when they are *supposed to*. (Dennett 1987, 290; emphasis in the original)

Now suppose one such vending machine is installed in Panama, where they use quarter-balboas, which are physically indistinguishable from quarter-dollars as far as the machine is concerned. So the two-bitser works correctly when accepting the quarter-balboas in this setting, though this would have counted as a mistake while the machine were located in the US. The question is how to identify the state the machine goes into when accepting balboas in Panama. An individualist would clearly say that no matter where it is located, the two-bitser goes into the same state Q; the only difference is its causal history. An anti-individualist suggests that while the *physical* state that the machine enters remains the same across the two environments, *intentionally* speaking the machine located in Panama goes into a different state—say, QB.

It is important to realize that unlike Burge, Dennett considers the choice between the two alternatives strictly speaking *indeterminate*. We can appeal to the fact that the vending machine is a functional device. It was designed by human engineers to serve certain purposes they had in mind. So there are some historical facts—facts about the origin—due to which the machine may be characterized as a device designed to give out soft drinks in exchange for the US quarters. If so, then also the state of the machine placed in Panama should be characterized in terms of the function it was selected for. With a reference to Millikan's biological definition of function (the details of which we shall discuss in section 4), Dennett claims that

> whether [the two-bitser's] Panamanian debut counts as going into state Q or state QB depends on whether, in its new niche, it was *selected for* its capacity to detect quarter-balboas—literally selected, e.g., by the holder of the Panamanian Pepsi-Cola franchise. If it was so selected, then even though its new proprietors might have forgotten to reset its counter, its first "perceptual" act would count as a correct identification by a q-balber, for that is what it would *now* be *for*. [...] If, on the other hand, the two-bitser was sent to Panama by mistake, or if it arrived by sheer coincidence, its debut would mean nothing, though its utility might soon—immediately—be recognized and esteemed by the relevant authorities [...], and thereupon its *subsequent* states would counts as tokens of QB. (Ibid., 293; emphasis in the original)

Dennett is confident that Burge and other intentional realists would agree that intentional ascription in the case of artefacts is a matter of practical expediency, or perspective, or stance. In the case of persons, however, these realists would insist that there was a *fact of the matter* whether someone meant aluminium, or arthritis, or whatever. But I suggest leaving the controversy over intentional realism for the following section.

Rather, let us now turn to the justification behind Dennett and Millikan's identification of representational states in terms of natural functions or purposes. This natural teleology is justified by that particular interpretation of evolutionary biology that I mentioned in the beginning of this chapter—namely, adaptationism. According to the definition that I quoted, adaptationism is the conviction that natural selection is only a significant source of the observed diversity of living forms. But let us introduce some distinctions here. Sober's definition closely corresponds to what Peter Godfrey-Smith calls *empirical adaptationism*, namely the claim that

> [n]atural selection is a powerful and ubiquitous force, and [...] [t]o a large degree, it is possible to predict and explain the outcome of evolutionary processes by attending only to the role played by selection. (Godfrey-Smith 2001, 336)

This should be distinguished from two stronger theses: *explanatory adaptationism*, according to which

> [t]he apparent design of organisms, and the relations of adaptedness between organisms and their environments, are the *big questions*, the amazing facts of biology [...] Natural selection is the key to solving these problems; selection is the *big answer* (ibid.; emphasis in the original),

and *methodological adaptationism*, which says that

> [t]he best way for scientists to approach biological systems is to look for features of adaptation and good design. Adaptation is a good "organizing concept" for evolutionary research. (Ibid., 337)

Godfrey-Smith argues that Dennett and Dawkins are explanatory adaptationists, sometimes even combining this—as when they marvel at the sheer *amount* of adaptive features in nature—with the empirical claim. I shall leave Dawkins aside, but I think that Dennett in particular actually subscribes to the strongest, i.e. methodological, adaptationism.

Consider such dramatic comments from Dennett's popular book, *Darwin's Dangerous Idea* (1995) as:

> Adaptationist reasoning is not optional: it is the heart and soul of evolutionary biology. Although it may be supplemented, and its flaws repaired, to think of displacing it from central position in biology is to imagine not just the downfall of Darwinism but the collapse of modern biochemistry and all the life sciences and medicine. (Dennett 1995, 238)

It seems that Dennett claims here that the assumption of good adaptedness is not just a correct answer to the key question of biology, but precisely a "good organizing principle" of all the life sciences, without which they would be unthinkable. I shall elaborate on this in a minute. But first, I need to take note of the fact that precisely the radical challenge to adaptationism that Dennett finds unthinkable arose in the midst of the biological science, in the famous paper by Stephen J. Gould and Richard Lewontin "The Spandrels of San Marco and the Panglossian Paradigm" (1978). In it, Gould and Lewontin deplore the assumption of

> the near omnipotence of natural selection in forging the best among possible worlds. This program regards natural selection as so powerful and the constraints upon it so few that direct production of adaptation through its operation becomes the primary cause of nearly all organic form, function, and behaviour. (Gould and Lewontin 1978, 76)

For Gould and Lewontin, many adaptationist explanations are unfalsifiable "just-so stories," and many alleged adaptations are mere "spandrels"—non-optimal by-products of a variety of constraints on natural selection. Godfrey-Smith argues Gould and Lewontin seek to undermine both empirical and methodological adaptationisms. Or, more precisely, they wish to uproot methodological adaptationism by depriving it of the support it gets from the alleged empirical evidence of good design.

Now back to Dennett's notion of adaptationism. Compared to Millikan who dismisses Gould's arguments, especially his rejection of adaptive character of cognitive capacities (see Millikan 1993, 46-47), Dennett is conciliatory. He interprets Gould and Lewontin's critique as a useful reminder that we should be careful, not hasty, adaptationists. But he rejects their suggestion to supplant adaptationism with the idea of *Baupläne*—the by now largely defunct theory that adaptation was good enough to explain certain superficial features of the design of organisms, but not their fundamental "body plans" (cf. Gould and Lewontin 1978, 85-89). Drawing on his distinction between "cranes" and "skyhooks" (cf. Dennett 1995, 73), Dennett asks what else than a mysterious skyhook could pull a complete body plan into existence, if the humble crane of natural selection were prohibited? Yet the most important point brought up by Dennett against Gould and Lewontin's critique is that they misunderstand the nature of adaptationism. The latter is not strictly speaking a theory. Only a theory—a collection of claims—could be either falsified or unfalsifiable. Rather, adaptationism is a stance that biologists are bound to adopt vis-à-vis the process of natural selection, lest they miss certain real patterns in nature.

The concept of a stance is, of course, an import from Dennett's philosophy of mind. Dennett argues that something is a bearer of representations only from the standpoint of an "intentional stance," which ascribes these representations under the constraint of an ideal rationality. As he puts it

> *all there is* to being a true believer is being a system whose behavior is reliably predictable via the intentional strategy, and hence *all there is* to really and truly believing that *p* (for any proposition *p*) is being an intentional system for which *p* occurs as a belief in the best (most predictive) interpretation. (Dennett 1987, 29; emphasis in the original)

Thus, the intentional stance licenses mentalism, while adaptationism licenses a sort of natural teleology. The two stances are analogous in that the former makes an assumption of rationality, whereas the latter that of optimality of design:

> When we adopt the intentional stance toward a person, we use an assumption of rationality or cognitive/conative optimality to structure our interpretation [...] In biology, the adaptationists assume optimality of design in the organisms they study. (Dennett 1990, 187; emphasis in the original)

Far from being expressions of naïve optimism either with respect to the rationality of agents or the optimality of organisms, the intentional stance and adaptationism are necessary presuppositions for answering certain "why"-questions. In the domain of psychology, we ask why an agent engaged in this or that behaviour; and we proceed by inquiring into what a perfect reasoner would do, given her circumstances; and in due time we are bound to discover that no agent is perfect in her reasoning. In biology, we ask why an organism was designed in a particular way; we hypothesize how it should be optimally designed, given what we know about its environment; and our prediction is going to be falsified (*pace* Gould and Lewontin, an adaptationist story that were too perfect would be useless to biologists, since it would teach them too little). And this is the gist of Dennett's position concerning adaptationism as a stance, which I think justifies its classification as a methodological, not just explanatory, thesis: "Adaptationism and mentalism [...] are not theories in one traditional sense. They are stances or strategies that serve to organize data, explain interrelations, and generate questions to ask Nature" (Dennett 1987, 265).

In view of the above, I think that Dennett's critics—among whom, as we shall see in the following section, we must count Burge—who ascribe to him the idea that stances are adopted or vacated arbitrarily or opportunistically are too quick.

It seems, then, that adaptationism is the basic premise of a Darwinian theory of the mind. As I already mentioned, adaptationism enables a sort of natural teleology. And once there is a room for natural purposes, we are entitled to posit states whose function it is to represent an external environment. Biological functions are identified in terms of a past performance of the traits that are so functionally defined. Hence, representational states, too, are identified in terms of what their predecessors were supposed to do. Finally, anti-individualism follows as a natural upshot of a theory that identifies all traits of an organism, including its representational states, in terms of its historical relations.

This, I take it, is the basic structure of a Darwinian argument justifying a version of anti-individualism shared by Dennett and Millikan. I shall return to some differences between the two philosophers' specific construal of this argument in a moment. At this point, it is clear that they agree in seeing anti-individualism as an almost trivial outcome of a psychology understood as an integral part of cognitive ethology. Dennett argues that cognitive ethology becomes anti-individualistic once freed from the legacy of *behaviourism*. He cites a particular research on vervet monkeys that have developed, in their natural habitat in the Sub-Saharan Africa, a relatively elaborate communication system involving several types of warning calls signifying the presence of different kinds of predators (see Cheney and Seyfarth 1990). Most of what is interesting about the lives of these animals would be simply missed from the perspective of behaviourism. For example, only from the intentional stance can we recognize a monkey as *deceiving*—i.e., as wishing to be believed as believing something that it does not. Certain fruitful hypotheses can be framed only based on the assumption that the vervets are intentional agents, i.e. having beliefs, intentions, fears and a host of other propositional states in the aetiology of their behaviour. Dennett argues that cognitive ethologists actually adopt the intentional stance towards the animals whose behaviour they study (see Dennett 1987, chap. 7 and Dennett 1998, chap. 20).

Now, the contrast between mentalism and behaviourism within ethology is important for the controversy over psychological anti-individualism in that behaviourism is a paradigmatically individualist doctrine. A behaviourist takes into account only the narrowly conceived inputs and outputs of behaviour, so that two individuals there were behaviourally identical would be psychologically identically as well. It is thus significant that Dennett stresses that, from the intentional stance, hypotheses about what the observed individuals believe and desire need to be framed "by figuring out what they ought to believe and desire, given their circumstances" (Dennett 1998, 292). Only within the particular circumstances of their natural environment could we figure out what, if anything, vervet monkeys' calls possibly mean, and we are licensed to attribute to these animals corresponding mental states.

Millikan also construes intentional psychology as part of cognitive ethology, but she sees the latter as haunted by the legacy of *individualism*, rather than behaviourism. She writes:

> Will a mature cognitive psychology need to characterize its subjects in ways that make reference to how they are imbedded in their environments? Or will it be "individualistic," making reference only to what supervenes on the structures of individual bodies and brains? The individualists argue that the behavioral dispositions of a person clearly depend only on that person's *inner* constitution, and hence that there can be no need to refer to the individual's relation to the wider environment in order to explain them. The anti-individualists argue that it is impossible even to describe much of the behavior that it is psychology's job to explain without reference to the environment. (Millikan 1993, 135; emphasis in the original)

As I pointed out in the previous paragraph, behaviourism is naturally interpreted as a type of individualism, but Millikan doesn't reject behaviourism *tout court*. She thinks individualistic and non-individualistic versions of behaviourism could be distinguished. She goes on to spell out the anti-individualistic notion of behaviour with the help of her concept of "proper function," to which we shall return in the next section. Here it suffices to say that the proper function of a trait is identified as that function the (by and large) successful performance of which enabled the trait's ancestors to copy or reproduce themselves in subsequent generations. A proper identification of a trait, including a behavioural trait, thus has a necessarily external element in the form of an *historical* dimension. But Millikan adds that the present, not just past, relations between behaviours and their environments are a necessary condition of a correct identification of behaviours. Millikan goes as far as suggesting that psychology, properly construed as an integral part of ethology, needs to construe its subject matter broadly, as involving both an organism narrowly conceived *and* its natural habitat.

> It is a very serious error to think of the subject of the study of psychology and ethology as a system spatially contained within the shell or skin of an organism. What is inside the shell or skin of the organism is only half of a system; the rest, if the organism is lucky, is in the environment. The organismic system, especially (indeed, by definition) the behavioral systems, reach into the environment

and are defined by what constitute proper, or normal, relations and interactions between structures in the organism and in the environment. (Ibid., 158)

Millikan's broad concept of behaviour is attractive and she seems to be careful enough not to extend her thesis into an implausible claim that mental representations themselves, let alone the mind, stretch into the external world. Burge himself argues for a broad construal of behaviour (see Burge 2007, 227), while deploring the tendency, popularized by some idiosyncratic anti-individualists, to "extend the mind" beyond the bounds of the body of an individual (Clark and Chalmers 1998).

Earlier, I announced that, despite a broad agreement between them, there are important differences between Dennett and Millikan's respective versions of a Darwinian theory of the mind. In the remainder of this section, I shall touch on two closely related points of difference.

Dennett seems to assume that the intentional stance is somehow basic in both psychology and biology. We start by making assumptions about what a rational agent would do or what an optimal design should look like before we inquire about the functional architecture of the agent or the evolutionary origin of an organism. By contrast, Millikan argues that the intentional stance needs to be underwritten by what Dennett calls the "design stance."

> where one ignores the actual (possibly messy) details of the physical constitution of an object, and, on the assumption that it has a certain design, predicts that it will behave *as it is designed to behave* under various circumstances. (Dennett 1987, 16-17; emphasis in the original)

For Millikan, the fact that something is at all interpretable from the intentional stance is evidence that it was designed: "There is nothing that exhibits apparently rational patterns for any time or in any detail that was not designed to do so, either by natural selection, or by something that natural selection designed" (Millikan 2000, 60). In his response to Millikan, Dennett concedes that the design stance *is* more basic "in the sense [Millikan] defends" (Dennett 2000, 341). That is, anything that is capable of a rich diversity and flexibility in its behavioural and perceptual responses is bound to have been designed either artificially or, ultimately, by means of natural selection. And yet, until a more principled way of distinguishing real intentional systems from merely apparent ones become available, Dennett says he prefers to keep his more "open-ended" approach that licenses the adoption of the intentional stance even toward simple artefacts (thermostats) or primitive organisms (frogs).

This last point ultimately rests on our two philosophers' divergent views of the reality of mental states. Millikan is a realist who believes that "folk psychology postulates inner items (for example, structures or events or states or entities)," and that "folk psychology is probably right" (Millikan 1993, chap. 3).

As for Dennett, he occasionally committed the tactical error of embracing the label "instrumentalist" to describe his theory of representational states (Dennett 1987, chaps. 2 and 3). Beside the fact that instrumentalism proves too difficult to distinguish from fictionalism, or just plain old anti-realism, despite Dennett's valiant effort (cf. Dennett 1987, 69-81), I think no single label is going to do justice to the complexity of his theory. It is true that some of his claims sound straightforwardly anti-realistic: "Folk psychology is *abstract* in that the beliefs and desires it attributes are not—or need not be—presumed to be intervening distinguishable states of an internal behavior-causing system" (Dennett 1987, 53). In other words, it is unlikely that scientific psychology will discover discrete items in the brain that corresponded to the beliefs and desires postulated by folk psychology. So Dennett suggests that we split folk psychology—which is a sort of a mixed bag, as it is couched in semantic terms, yet also postulating a particular sort of entities—into two new theories. On the one hand, there would be the "intentional systems theory"—i.e., an abstract science of rationality, akin to decision theory or game theory—and the "sub-personal cognitive psychology"—i.e., a concrete science of the neural systems. The former would be dealing in pure semantics, the latter in pure syntax. I take it that the construal of the intentional systems theory as a purely abstract theory is supposed to guarantee the metaphysical sanity of Dennett's willingness to attribute intentional states to natural selection itself. Millikan probably reads her own intentional realism into Dennett's theory when she finds the talk of beliefs and reasoning of Mother Nature "otiose in biology" (Millikan 2000, 65). She elaborates that "there is no sense in such talk *because there is nothing in Nature analogous to beliefs and nothing that so much as reminds one of inference*" (ibid., 64; emphasis in the original). I think Dennett's response to this is ingenious:

> As Sherlock Holmes, the patron saint of inference, famously said, once you have eliminated all other possibilities, the one that remains, however improbable, must be the truth. Is that not an inference? Does not Mother Nature eliminate all other possibilities, on a vast (not actually Vast) scale, thereby "inferring" the best design? When Deep Blue eliminates a few billion legal moves and comes to rest on one brilliant continuation, it surely reminds Kasparov of inference! (Dennett 2000, 343)

While I am convinced that Dennett's attribution of intentions to natural selection is not metaphysically weird for the reasons offered by Millikan, I *do* find a metaphysical difficulty within Dennett's position. I shall elaborate on it in the following section.

3 Deflationism and Realism

Burge rejects Dennett's construal of mental representation. He finds this view not only implausible, but obviously so: "I mention [it] only to lay it aside" (Burge 2010, 293). He dubs it the "deflationary view" of representation:

> On this view, treating something as engaging in representation is merely a matter of a "stance," with more or less practical or instrumental value. On such a position, there is no objective kind, *representation*, that can be discovered through normal scientific investigation. On such a position, there is no more *theoretical* reason to treat an individual as having beliefs or perceptions than there is to treat a vending machine, or a planetary system, as representing something. It is all a matter of practical convenience or optional attitude toward the phenomena. (Burge 2010, 293; emphasis in the original)

Dennett returns the favour by dismissing Burge's position as one more example of the traditional belief in the "original intentionality" of human minds, from which all other intentionality—ascribable, as the case may be, to the artefacts of our own design, or to the inanimate objects of nature—is "derived."

Superficially at least, Burge unites with Millikan against Dennett with respect to the issue of the reality of representational states. Burge and Millikan are intentional realists while Dennett is a sort of anti-realist about the mind. Ultimately, however, Burge is going to classify Millikan's theory, despite its realism, as another variant of the sort of naturalism of which an anti-realist version is Dennett's view. Millikan as well as Dennett turn out to be equally unacceptable to Burge as two models of a basically reductionist view of the representational mind. Yet the arguments on both sides are subtle. I propose, first, to rehearse Dennett's reasons, derived from his understanding of Darwinism, for intentional anti-realism; second, I examine Burge's grounds for rejecting both Dennett and Millikan's naturalistic theories.

Recall that Dennett starts motivating his anti-realism by construing the thought experiment placing a lowly artefact, not a person, in two different environments. Dennett expects that everybody is going to agree that representational states cannot be attributed to mere artefacts literally, so that an uncertainty as to whether to attribute one state rather than another is not disquieting. Next, Dennett needs to demonstrate that a similar indeterminacy befalls intentional description of persons as well. In other words, he needs to show that in the case of persons as well as artefacts, there is no bedrock fact when it comes to possessing representational states, but a mere useful way of speaking. For that purpose, Dennett offers an additional thought experiment.

Suppose someone decided to survive into the twenty-fifth century in a hibernation device of some sort. He would be wise to make that device mobile, so that it can look for the sources of energy. And since these are bound to be scarce, the mobile hibernation device should be capable of fighting off the machines of other people who—as it might be expected—would build survival machines of their own. The more sophisticated such machines get, the better their chance to deliver their hosts into the future. Hence we might expect the best machines would be robots capable of self-control, of setting their own goals based on their assessment of a current situation, and so on. Now, the intentional realists such as Burge would, according to Dennett, insist that such robots, no matter how sophisticated, have whatever fake intentionality they possess, ultimately derived from our plans and purposes. But here is the clincher: "the conclusion forced upon us is that our own intentionality is exactly like that of the robot, for the science-fiction tale that I have told is not new; it is just a variation of Dawkins' [*The Selfish Gene*] vision of us [...] as 'survival machines' designed to prolong the futures of our selfish genes" (ibid., 298). Where Dennett's story started with the real meaners as the ultimate source of design, it turns out those meaners mean no more literally than the selfish genes of Dawkins's story. And yet, Darwinism shows that we can get intelligent design without any real minds: "when natural selection selects, it can 'choose' a particular design *for one reason rather than another*, without ever consciously—or unconsciously!—'representing' either the choice or the reasons" (ibid., 299, emphasis in the original). Hence we see again that not only can we attribute intentions to Mother Nature, despite the fact that she is no real reasoner, but a mere process of natural selection—but we do the same with persons, despite the fact that, strictly speaking, *they* are no real reasoners, either. There are only *ersatz* thinkers, but anything can be considered as such, if selected properly either artificially or naturally.

This is how it works according to Dennett. We attribute beliefs, desires and other attitudes to each other, but there is no way these folk psychological states—imagined to be both semantic and holistic, as well as concrete and discrete, entities—are going to be recognized by a mature science of psychology. Therefore, Dennett suggests splitting folk psychology into two new theories: "one strictly abstract, idealizing, holistic, instrumentalistic—pure intentional system theory—and the other a concrete, microtheoretical science of the actual realization of those intentional systems—what I will call sub-personal cognitive psychology" (Dennett 1987, 57). At the intentional system level, we are semantic engines; at the microtheoretical level, we are physical, or perhaps syntactic systems. How do these two levels of description relate to each other? In other words, how does the brain, a mere syntactic engine, produce semantics? Dennett answers:

> It cannot be designed to do an impossible task, but it could be designed to *approximate* the impossible tasks, to *mimic* the behavior of the impossible object (the semantic engine) by capitalizing on close (close enough) fortuitous correspondences between structural regularities—of the environment and of its own internal states and operations—and semantics types. (Ibid., 61, emphasis in the original)

As far as I can see, Dennett suggests that the brain behaves *as if* it were a semantic engine, in addition to being a syntactic one. We don't have an ability to build brains (yet), but we can build much simpler devices that fulfill some semantically characterizable tasks. For instance, we could build a machine that would catch the telephone communications that are death threats, by picking out words like "... I will kill you..." or "... you... die... unless..." and such (cf. ibid., 62). If so, we would succeed in building a "death-threat interceptor"—that is, a purely syntactic device which is also describable in such semantic terms. The machine would be primitive and unreliable, but we could keep on improving it.

This much could be achieved by artificial design, but what about natural selection? Dennett claims that our brains are dumb syntactic devices that were selected for their ability to mimic semantic engines, and have kept on getting better at this over time: "in the end all one can hope to produce (all natural selection can have produced) are systems that *seem* to discriminate meanings by actually discriminating things (tokens of no doubt wildly disjunctive types) that co-vary reliably with meanings" (ibid., 63; emphasis in the original). So we can interpret each other intentionally owing to a long history of a (more or less) successful coping of our species with its environment. It is due *solely* to the benefit of hindsight afforded by this history that we can *appear* to be reasoners and meaners.

Burge dismisses this whole approach because he disagrees that the sort of responsiveness to stimuli that could be found in nearly all living things captures the kind *representation* employed in psychological explanation. In his critique of Dennett, Burge makes a point he has repeated in polemics with many naturalists over the decades. The point is that these authors understand the relation between science and metaphysics backwards. Their projects are driven by various metaphysical interests, in particular by the interest to make representation and the mind non-mysterious. For example, one worry that seems to have motivated a lot of attempts in the past few decades to naturalize intentionality is epiphenomenalism. Many philosophers assumed that all the causal work is done by the underlying physical processes, while representations *qua* representations are causally inert. But, according to Burge, we should eschew such preconceptions of a (materialistic) metaphysic and instead begin with studying the actual practice of psychological explanation in which intentional idioms figure prominently. Such explanation works in everyday contexts and it is part of a mature scientific psychology as well. Last but not least, it is central to our self-image as agents (see Burge 2007, chap. 16). Hence, there is nothing *prima facie* mysterious about mental representation as it figures in a successful, testable and precise psychological explanation.

When it comes to Dennett, it seems that his project of replacing folk psychological concepts—despite the fact they are commonplace in research programmes of perceptual psychology and elsewhere—with the two new theories of abstract and sub-personal psychology, respectively, is *stipulative*. It is driven by a metaphysical worry that brains do not possess semantic properties. And Dennett's conclusions are unclear. He doesn't seem to make up his mind as to how seriously should the talk of mental states be taken. On the one hand, when he says that brains "mimic" semantic engines, or that intentional ascription is indeterminate, representations seem merely useful fictions. On the other hand, he also suggests that brains "realize" semantic states (Dennett 1987, 67). The relation of realization, familiar from the functionalist literature, seems more robust than mere "mimicking." Yet Dennett does not elucidate how similar or different these two relations are.

Burge raises a similar charge of stipulativeness against Millikan's account of representation: "I believe that Millikan's view amounts to a stipulation about how she intends to use 'representation'" (Burge 2010, 300). He does praise her account for separating representation from mere information-carrying. The latter is straightforwardly causal, so that an organism goes into a particular information-carrying state whenever the appropriate causal prompt is present. There is no room for misrepresentation or mistake. We saw how that room is created by Millikan's appeal to selectionist history. A state of an organism can *mis*represent, if there is a norm set by the past performance of the ancestors of that state, since then we can say how the state is supposed to function, even though it actually doesn't. There is nothing in this notion that precludes its ascription even to artefacts, as we saw in Dennett's case. The same phenomenon could be described just by using the notion of biological function, normal environmental conditions and sensory discrimination.

4 Historical Function, Systems Function, and Individualism

In this last section, I should like to focus on the concept of function appealed to by Dennett and Millikan. It should become clear that this concept motivates a version of anti-individualism, which is actually incompatible with Burge's original theory. Which is another way of saying that Dennett and Millikan's respective construals of anti-individualism differ from Burge's. It is then surprising to find even Burge, as he does in his defence of the autonomy of psychology, to the biofunctional concept of function, because it results in an inconsistency.

As is well-known, there are two main concepts of function: Millikan's historical theory of function and Robert Cummins's systems theory. I have already explained some elements of Millikan's theory earlier in this chapter. A comparison with Cummins's view might help further to clarify the nature of Millikan's view.

Cummins picks on the complexity of systems of various sorts and ascribes functions to such systems on the basis of the workings on their parts. To be more precise, an item x has a function φ within a system s, assuming a background of an analytic explanation of x, which appeals to x's capacity to φ in s. Cummins uses an example of the heart to illustrate his proposal: "It is appropriate to say that the heart functions as a pump against the background of an analysis of the circulatory system's capacity to transport food, oxygen, wastes, and so on, which appeals to the fact that a heart is capable of pumping" (Cummins 1975, 64). Although this example is taken from biology, notice that Cummins could apply his approach in assigning functions to inanimate systems as well–thus, our solar system could be regarded as a functional system. Also notice that Cummins refers only to the current properties of a functional item, and that he confines his attention to the internal parts of a system.

Curiously, Millikan also illustrates her alternative theory of function with the example of the heart. Let me quote a relevant passage in its entirety:

> A heart, for example, may be large or small (elephant or mouse), three-chambered or four-chambered, etc., and it may *also* be diseased or malformed or excised from the body that once contained it, hence unable to pump blood. It falls in the category *heart*, first, because it was produced by mechanisms that have proliferated during their evolutionary history in part because they were producing items that managed to circulate blood efficiently in the species that contained them, thus aiding the proliferation of that species. It is a *heart*, second, because it was produced by such mechanisms in accordance with an explanation that approximated, to some undefined degree, a Normal explanation for production of such items in that species and bears, as a result, some resemblance to Normal hearts of that species. By a "Normal explanation" I mean the sort of explanation that historically accounts for production of the majority of Normal hearts of that species. And by a "normal heart," I mean a heart that matches, in relevant respects, the majority of hearts that, during the history of that species, managed to pump blood efficiently enough to aid survival and reproduction. (Millikan 1993, 55)

Although Millikan agrees with Cummins in assigning the same function of blood-pumping to the heart, the rationale is importantly different. A particular exemplar of the heart has the function that it does in virtue of its ancestry. There has been a long line of organs that more often than not succeeded in pumping blood in the past and the present exemplar is their descendant. Therefore, even if our particular exemplar fails to pump blood efficiently, or even if it is so defective as to never having pumped any blood, we can still correctly identify it as the kind *heart* in virtue of its relation to the line of ancestral hearts that have enabled the survival up to now. In other words, we can assign a function properly in virtue of a background of normality. (That is why Millikan speaks of "normal explanation").

We can see now that the historical concept of function is narrower than the systems view, in that the former is restricted to items that are products of design of one sort or another, whereas the latter was applicable to any complex system, whatever its origin. In another sense, however, Millikan's concept of function is obviously broader than Cummins's concept. The historical function is identified as such in relation to things outside of the system of which it is a part of, including things in a distant past. By contrast, the Cummins function disregards relations between a functional system and its surroundings or its origin.

We have already seen in section 2 that the historical account seems well suited for the purposes of establishing anti-individualism, since it enables us to run the familiar thought experiments. Dennett's thought experiment featuring a vending machine whose powers to detect currency are affected by an environment directly draws on Millikan's construal of functional ascription. In section 3, I explained that Burge does not like Dennett's conclusion that functional, hence semantic, ascription remains forever indeterminate. But Burge does not like Millikan's account of representation, despite its intentional realism, either. I showed that Burge sees these naturalist theories as two versions of "deflationism," namely a tendency to stipulate various minimal detection capacities in place of the robust concept of representation, which is at home in everyday life and scientific psychology. Recall that for Burge, psychology types its kinds anti-individualistically, yet there is no need to reduce it to some lower, presumably individualistic, level of discourse.

In view of this critique of deflationism, it is then surprising to find Burge supporting his view of psychology by means of a twin story featuring the biological item that we have already seen in both Cummins and Millikan—the heart. Burge invites us to imagine a physical replica of the human heart placed in an alien body:

> Something is a heart because its organic function is to pump blood in a circulatory system that extends beyond the surfaces of the heart. One can imagine an organ in a different sort of body with a totally different function (it might pump waste for example). The causal powers attributed to such an organ by biology would be different from those attributed to a heart. Such an organ would not be a heart, but it might be chemically and structurally homologous to a heart. (Burge 2007, 323)

Like Millikan, Burge is explicit that in order to categorize properly the physiological kind *heart*, we must attend to something external to its instantiations—namely, their selectionist history:

> To be a heart, an entity has to have the normal, *evolved* function of pumping blood in a body's circulatory system. One can conceive of a physically homologous organ whose function is to pump waste—or even a physically homologous entity that came together accidentally and lacks a function. Such entities would not be hearts. (Ibid., 326; emphasis added)

Thus, in his claim that the heart and the alien waste-pump differ in terms of their divergent histories, Burge assumes the historical notion of function.

The point of presenting physiology as another special science that types its kinds anti-individualistically is to suggest that intentional psychology is no worse. If physiology enjoys respectability at least comparable to that of physics, then psychology should not be looked down upon, either. In accordance with his methodological decision to prioritize an actual scientific practice over metaphysics, Burge seems to be suggesting that the claims of naturalism are satisfied by taking a successful explanatory practice in psychology and other special sciences at its face value. There is no call for trying to force that practice into a straightjacket of some reductive metaphysics.

However, I wonder whether the above defence of the respectability of intentional psychology is entirely consistent with Burge's anti-individualism. It is true that Millikan's historical concept of function, exploited by both Burge and Dennett, appears to be anti-individualistic in character compare to Cummins's theory. For Cummins, who takes into account only the internal parameters of a functional item, there is no way to distinguish between a blood-pump and a waste-pump. In a parallel case, an individualist in psychology has no way of distinguishing between aluminium-thought and twin aluminium-thought as long as he restricts his attention to internal parameters of a thinker alternating between two environments. Millikan provides a resource for drawing the required distinction. A heart is distinct from a waste-pump in terms of its divergent evolutionary origin.

Yet recall Burge's recent definition of anti-individualism that I quoted at the outset. It is meant as a theory of the very *nature* of representational mental states. That means it is not merely a theory of *description* of these states. It does not merely say that in speaking of representations, we should mention their relations to an environment. Dennett could be easily critiqued as misinterpreting anti-individualism as a descriptive, rather than metaphysical, theory. In his thought experiment about the vending machine, we saw him making an explicit assumption that the machine goes into a particular state that could be described differently relative to an environment. If so, the nature of such a state should be identifiable independently of its various environmental descriptions. In fact, Dennett in at least one text admits as much. For evidence of Dennett's betrayal of externalism, see his response to Frank Jackson: "So let me confirm Jackson's surmise that I am a behaviorist; I unhesitatingly endorse the claim that 'necessarily, if two organisms are behaviorally alike, they are psychologically exactly alike'" (Dennett 1993, 923). Vending machines certainly behave the same: accepting coins and churning out bottles of soft drink. So some narrow description of what they represent must in principle be available, too. The description of machines which takes into account such facts as their location in the US or Panama, respectively, is something extraneous. And this might perhaps help to find a correlate of environmentally identified state at the sub-personal, or syntactic, level.

Now, the above critique of Dennett as a closet individualist, or perhaps someone who just misunderstood anti-individualism, can unfortunately be extended to Burge's own theorizing about the autonomy of intentional psychology. It is crucial to realize that there is a narrow, individualistic way of identifying both the heart and its alien counterpart: they are both a kind of *pump*. If so, an environmental description of this organ is strictly speaking optional, since we can descend to a lower, individualistic level. It is true that Burge is speaking of biology (or perhaps physiology), but the point of his example clearly is that psychology is analogous to biology. In section 1, I quoted Burge putting the heart example to a somewhat different use—namely, arguing that this organ does not consist of its external relations—but even here, he said: "I think that representational mind is like that" (Burge 2010, 66). So, Burge appears to take an analogy between psychology and biology very seriously. And yet, elsewhere he opposed attempts to dilute anti-individualism to a mere theory of description in terms of environmental relations. He said that in lower-level science, we often do have alternative ways of identifying the instances of explanatory kinds. In psychology, this is not available. "We have no such ways of identifying states of the body that (putatively) are beliefs, independently of assumptions about the beliefs" (Burge 2007, 353). Accordingly, though both the heart and an alien organ belong to the kind pump, there is no way to identify a thought individualistically.

I conclude that both naturalistic attempts to found representation ultimately in evolutionary biology and Burge's nonreductive attempt to preserve the autonomy of intentional psychology are ridden with problems. Burge may be right that what Dennett or Millikan succeeded in deriving from biology is not representation, as it is understood in everyday life and psychology, but something too minimalistic. On the other hand, Burge, despite his claim that a widespread fear of reductive metaphysics is simply a prejudice, still seems wishing to connect psychology with biology, which entangles his account in an inconsistency.

References

Burge, T. (2007). *Foundations of Mind*. Oxford: Oxford University Press.
_____ (2010). *Origins of Objectivity*. Oxford: Oxford University Press.
Cheney, D. L. and Seyfarth, R. M. (1990). *How Monkeys See the World*. Chicago and London: The University of Chicago Press.
Clark, A. and Chalmers, D. (1998). The Extended Mind. *Analysis* 58, (1): 10-23
Cummins, R. (1975). Functional Analysis. In Sober 1994: 49-69.
Dennett, D. (1987). *The Intentional Stance*. Cambridge, Mass.: The MIT Press.
_____ (1990). The Interpretation of Texts, People and Other Artifacts. *Philosophy and Phenomenological Research* 50, Supplement, Fall, 177-194.

_____ (1993). The Message Is: There Is No *Medium*. *Philosophy and Phenomenological Research* 58 (4): 919-931.

_____ (1995). *Darwin's Dangerous Idea*. New York: Simon & Schuster.

_____ (1998). *Brainchildren*. Cambridge, Mass.: The MIT Press.

_____ (2000). With a Little Help from My Friends. In Ross, Brook and Thompson 2000: 327-388.

Godfrey-Smith, P. (2001). Three Kinds of Adaptationism. In *Adaptationism and Optimality*. S. H. Orzack and E. Sober (eds.). Cambridge: Cambridge University Press, 335-357.

Gould, S. J and R. C. Lewontin (1978). The Spandrels of San Marco and the Panglossian Paradigm: A Critique of the Adaptationist Programme. In Sober (1994): 73-90.

Millikan, R. (1993). *White Queen Psychology and Other Essays for Alice*. Cambridge, Mass.: The MIT Press.

_____ (2000). Reading Mother Nature's Mind. In Ross, Brook and Thompson (2000), 55-75.

Ross, D., A. Brook and Thompson, D. (eds.) (2000). *Dennett's Philosophy: A Comprehensive Assessment*. Cambridge, Mass.: The MIT Press.

Sober, E. (ed.) (1994). *Conceptual Issues in Evolutionary Biology*. 2nd Ed. Cambridge, Mass.: The MIT Press.

_____ (1996). Six Sayings about Adaptationism. In *The Philosophy of Biology*, D. L. Hull and M. Ruse (eds.) Oxford: Oxford University Press 1998, 72-86.

14

Creatures of Norms as Uncanny Niche Constructors[1]

Jaroslav Peregrin

1 Introduction

Imagine a Paleolithic hunter who has failed to hunt down anything for a couple of days and is hungry. He has an urgent desire, the desire to eat, which he is unable to fulfill – his desire is frustrated by the world. Now imagine our contemporary bank clerk who went to work without his wallet and is hungry too. He too is unable to fulfill his urgent desire to eat because it is frustrated by the world.

From the viewpoint of the two individuals the situations are similar. However, there is at least one crucial difference. While the hunter cannot eat because there is no food in the vicinity (at least as far as he is aware), the clerk can easily get hold of tons of food - it would suffice to visit the nearest supermarket. The reason he cannot get the food is not that it is *physically impossible*, but because taking food from a store's shelves without paying is *forbidden*.

This story brings home the fact that many of the barriers that constrain our present lives, restricting us to paths only within the space to which they limit us, are no longer barriers in the literal sense of the word - they are no longer produced entirely by the conspiracy of the causal laws that form our physical niche. Rather, they are produced by the conspiracy of attitudes of our fellow humans—they are deliberate *rules*, rather than inexorable *natural laws*. In this way evolution is now canalized less by the physical environment relatively independent of it, and more by the ploy of the organisms it brought into being.

A full appreciation of this autocatalytic situation may lead us to a deeper understanding of certain philosophical doctrines, pervasive especially after Kant, regarding normativity as the hallmark of the human. We come to see how these doctrines get enlightened by scientific theories regarding the development of the human race and its continuities/discontinuities with its animal cousins.

---

[1] Work on this chapter was supported by research grant No. P401/10/0146 of the Czech Science Foundation.

2 Niche Construction as a Factor of Gene-Culture Co-evolution

Evolution, the popular wisdom says, is about the ways organisms adapt to the environment in which they live, the ways they utilize its sources and avoid its dangers. Hence the whole process can be depicted as a chain of reactions to, or perhaps the overcoming of, a "mismatch"—a mismatch between organisms (considered as a lineage), on the one hand, and their environment, on the other. In general, to do away with a mismatch between two entities, we might think of adjusting either of them (or, of course, both of them simultaneously); but as it is only organisms that evolution can directly "control," there would appear to be an asymmetry; the evolution appears to be restricted to manipulating the organisms. This is why the situation is usually seen as a one-sided adaptation.

But this is not the whole truth. Evolution *can* manipulate the environment, though of course only *via* the organisms living in it. Suppose there is a feature of an environment that crucially menaces the survival of the organisms (it may be some kind of predator, poisonous plant, or landscape feature, such as a hidden chasm etc.). Selection can respond to this situation not only by adapting the organisms so as to skillfully avoid this menace, but also by adapting them so that they themselves eradicate the menacing factors, thereby effecting a change in the environment itself. This is something that has been addressed, in the literature, under the heading *of niche construction* (Odling-Smee 1996, Laland *et al.* 2000, Odling-Smee *et al.* 2003). The authors often point out that this is a neglected way evolution works.

Of course, changing the organisms so as to avoid—or, alternatively, to exploit—some environmental feature would usually be an infinitely simpler task for evolution than making them permanently modify the environment. This is an obvious reason for the tendency to take a one-sided view of how evolution deals with an organisms-environment mismatch. But we should keep in mind that this does not hold unexceptionally— organisms clearly do change their environment in various more or less permanent ways; hence niche construction is not impossible.

The broader context in which the idea of niche construction is usually discussed in the literature is that of the so-called *gene-culture coevolution* (Cavali-Sforza and Feldman 1983, Durham 1991, Feldman and Laland 1996). The basic idea is that from the viewpoint of evolution, culture is not a mere by-product, an idle super-structure, of the genetic development, but rather its effective factor. Though culture, without doubt, is a product of genetic evolution, it firms up into a source of additional, paragenetic "inheritance," which, however, works inextricably from its genetic substrate.

In this context, niche construction is seen as one of the important tools of cultural inheritance. As Laland *et al.* (2000) put it:

Creatures of Norms as Uncanny Niche Constructors   191

We are suggesting that our ancestors constructed niches in which it "paid" them to transmit more information to their offspring. The more an organism controls and regulates its environment, and the environment of its offspring, the greater should be the advantage of transmitting cultural information from parent to offspring.

This leads the authors to the following picture:

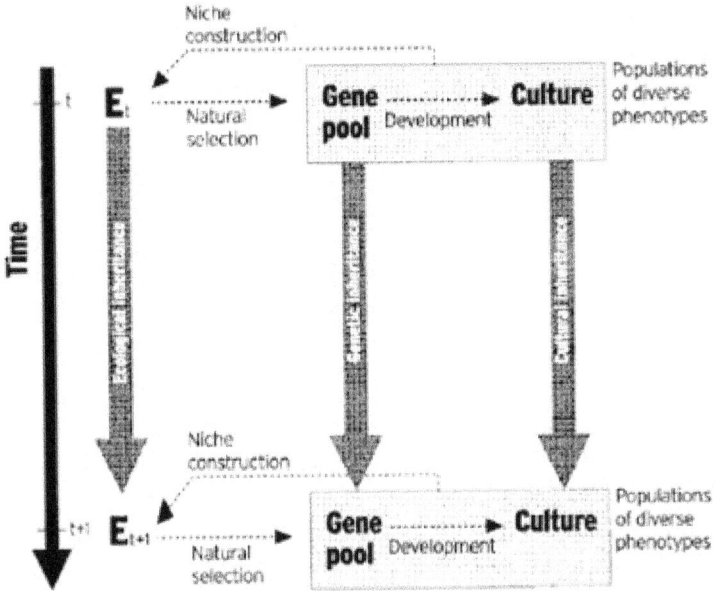

Laland *et al.* (ibid.) comment on the idea this picture is intended to capture as follows:

> It is possible that, once started, vertical cultural transmission may become an autocatalytic process: greater culturally generated environmental regulation leading to increasing homogeneity of environment as experienced by parent and offspring, favouring further vertical transmission. With new cultural traits responding to, or building on, earlier cultural traditions, niche construction sets the scene for an accumulatory culture.

[...]

> In the presence of niche construction, adaptation ceases to be a one-way process, exclusively a response to environmentally imposed problems; it becomes instead a two-way process, with populations of organisms setting as well as solving problems.

The authors also think that once we appreciate the importance of niche construction for cultural transmission, we will be able to see that the chasm between "nature and nurture" is not so vast. Though culture, as developed by us humans, is surely a quantitatively unprecedented phenomenon, it is not qualitatively different from what we find in non-human species; it amounts to a mere unprecedented inflation of tendencies that were already extant:

> Humans may be unique in their extraordinary capacity for culture, but they are not unique in their capacity to modify natural selection pressures in their environments. [...] Human culture may allow humans to modify and construct their niches, with spectacular ecological and evolutionary consequences, but niche construction is both general and pervasive and probably influences the ecology and evolution of many species.
>
> [...]
>
> Culturally modified selection pressures are now regarded not as unique, but simply as part of a more general legacy of modified natural selection pressures bequeathed by human ancestors to their descendants.

This indicates that, for the authors, the idea of niche construction contributes to underwriting the continuity between us humans and our animal cousins. In this chapter, I wish to emphasize another aspect which actually underwrites the *discontinuity* between humans and the rest of the animate word. I want to point out that the kind of niche construction that is crucial to us in our present state of development is no longer restricted to an elaboration of the *physical* world. It has surged forward into erecting what can be seen as *virtual* worlds, the anchoring of which within the physical one is only very loose.

3 Three Ways of Constructing One's Niche

When speaking of "niche construction," we must draw some important distinctions. In particular, we must distinguish several levels at which an organism may modify its niche in a way relevant for its further development. Let us discern three such levels (without claiming to rule out other ways of categorizing the spectrum).

1. *Modification of the natural physical niche*. Every organism, however primitive, changes its environment and it is likely that at least some such changes produce a feedback in the sense of modifying the selection pressures on the organisms in question. Such changes become significant for evolution when they are sufficiently large and systematic. (It is one thing to cope with an unfavourable environmental factor by means of evolving some kind of countermeasure; it becomes quite another if this factor can be eradicated, thus removing

the necessity to react to it.) This kind of changes clearly does underwrite the continuity between humans and other animal species w.r.t. niche construction, urged by Laland *et al.* (ibid.)

2. *Building an artificial physical niche atop the natural one.* Once the abilities of dealing with the environment surpass a certain boundary, an organism may become capable of literally building its own, artificial niche atop of the natural one. This undoubtedly largely changes the character of the pressures the organism faces. (Once you have a house where you can hide from night predators, you can forego the instincts that were vital for your survival when you slept in the open.)

Especially interesting artifacts, then, are those which can be deemed "symbolic," artifacts that *represent* the knowledge accumulated by preceding and current generations and permit it to be passed on to the next generation. Hutchins and Hazelhurst (1992) write:

> [C]ulture involves the creation of representations of the world that move within and among individuals. This heavy traffic in representations is one of the most fundamental characteristics of human mental life, yet since it is a phenomenon not entirely contained in any individual, it has largely been ignored by cognitive science. If each individual is capable of learning something about the environmental regularity and then *representing* what has been learned in a form that can be used by other individuals to facilitate their learning, knowledge about the regularity could accumulate over time, and across generations.

How these "symbolic artifacts" (books, in the most developed form) foster extragenetic, cultural evolution is obvious.

Clark stressed "our amazing capacities to create and maintain a variety of special external structures (symbolic and social-institutional)" (Clark 1997, 179), leading to the state where "intelligent brains *actively* structure their own external (physical and social) worlds so as to make for successful actions with less individual computation" (ibid., 191). In this way, we can say, we come to unload parts of our minds into the environment thus "supersizing our minds" (Clark 2008).

3. *Virtual niches.* Human societies not only elaborate and rebuild the physical niches in which they live and enhance them with symbolic artifacts; they also generate a brand new kind of pressures capable of channeling natural selection. This was noted already by Alexander (1989): as he put it, humans came to become "their own principle hostile force of nature." The fact that niche construction goes beyond the regulation of the forces of nature towards a "social niche construction" producing a certain kind of virtual environments superimposed upon the physical one has been tabled recently by a number of authors (Flinn, Geary and Ward 2005, Boyd and Richerson 2008, Odling-Smee and Laland 2009).

As we have indicated at the start of this article, people are diverted from certain actions not only because they *cannot* carry them out, due to the boundaries of their physical world, but also because they are *not allowed* to carry them out, due to normative barriers erected by their society. This is perhaps the most important, though largely neglected aspect of human niche construction.

There is also, I am convinced, a sense in which this fabrication of virtual, normative niches must precede the ability to produce "symbolic artifacts." To build a "symbolic artifact," such as a book, we need the stuff it is made of, namely symbols—in the typical case, a language. Now there are arguments (I have presented them elsewhere) for the claim that any symbolic system is constituted in terms of rules, i.e., precisely of the same kind of entities that are the scaffolding of our virtual niches. The trouble, it seems to me, is that the nature of this virtual space is currently not well understood.

Concepts usually employed to characterize the way in which a virtual niche influences the development of individuals include *imitation, transmission across generations* or *ecological inheritance*. I think that these concern merely one aspect of the situation, while a crucially important aspect goes almost unnoticed. What puts cultural inheritance into motion in the first place, and what continues to underlie it, is a specific kind of self-perpetuating behavioral (meta-)pattern that provides for virtual boundaries analogous to the tangible ones, as illustrated at the start of this article.

4 Virtual Niches

The characterizing aspect of our general human niche is that it is not constituted merely by inanimate objects and by the individuals of other species; but also by our conspecifics. And it is important to see that already this provides for a peculiar kind of niche construction: if the relevant environment of an organism is partly constituted by other organisms of the same kind, and if evolution always manipulates with the whole kind, then it changes the environment of the organism simply by means of manipulating the kind.

Within evolution theory, this has led to the employment of evolutionary game theory (Maynard Smith 1982). If an organism is not merely to react to the state of an environment that is independent of it in the sense that it is not influenced by the evolution of the organism, but also to other organisms that in turn react by means of their own evolution, then the "decisions" taken by evolution must assume the form of certain equilibria rather than of simple optimalization of features.

Hence considering the organism-environment relationship, we have *two* potential kinds of feedback: first, the part of an organism's environment that is constituted by other organisms (its conspecifics and perhaps organisms of some other species) reacts to the organism's evolution because it is subject to the same evolution; and the rest of the environment may "react" to the organism's evolution in that it is cultivated by the organism. Now I want to point out that our virtual niches may be seen as resulting from an interaction of these two feedback loops.

We humans not only try to outsmart each other in the battle over resources, we cooperate, share and jointly mine the resources, to an extent that has no precedent in any other species. (Of course, cooperation can be seen as merely a more sophisticated form of outsmarting everybody else; this, however, does not alter the fact that it is an uprecedented strategy.) A straightforward and commonly accepted explanation of the emergence of such a large-scale cooperation (and together with it what Boyd and Richerson [1998] call human "ultrasociality") is still awaited, though it would seem that many of the presumed ingredients for it have been well scrutinized (see, e.g., Nowak 2006 or West *et al.* 2007). Nevertheless, the fact that the cooperation happens is obvious.

Now cooperation is, from the very beginning, a matter of *rules* (though not necessarily the full-fledged, outspoken rules that fibrillate our advanced societies). To cooperate is to suppress one's immediate subjective needs (which hopefully gets rewarded in the long run) and to do so in tune with other individuals; hence what is needed is at least regularity across persons. However, for the cooperation to pervade and to graft into the next generation regularity is not enough; it must be regularity generally understood as something that *ought to be*—as something one *should* sustain and to which one should make other people, especially one's own offspring, conform.

Hence, as I argued elsewhere (Peregrin 2010), the emergence of rules as the entering wedge to (not only) cooperation, presupposes the ability of thinking in the "normative mode," of being able to understand that something that *ought to be*. And once this ability is in place, we have the resources for erecting virtual barriers (by means of what *ought not to be*) and by means of them to erect virtual worlds. Rules (though originally, as Sellars 1949, 299 put it, "written in nerve and sinew rather than pen and ink" and only later capable of being explicitly articulated) form the virtual barriers that are able to restrain members of human communities analogously to the tangible physical barriers restraining any inhabitant of the physical world.

Consider an example given by Joseph Heath (2008, 153):

> Most people, for example, when getting on to a bus, would like to sit down. Even if all the seats are taken, it is still possible to sit on the floor, or on someone's lap. One could simply order another person out of his seat, or request that he moves, or physically grab him and pull him out. Most people never even consider these options, simply because such behavior is inappropriate in the context. Instead, they will often give up their seats to persons more in need of them. They will also hesitate before taking a newly vacated seat, to see if anyone else is moving for it, so that they may seem duly deferential to the needs of others. All

of these constraints on the pursuit of one's objectives are a consequence of the set of social norms that govern social interactions on crowded buses (differentiated by age, gender, infirmity, and so on).

This duly illustrates the spontaneity with which we usually respect constraints implied by social norms: if we were asked where we can sit on a bus, the possibility of sitting on a place already occupied by another person (though it might be physically possible to remove the person from the seat and sit there) comes to us as similarly nonexistent as the possibility of sitting on a seat that is not physically there.

In this sense, the normative constraints yielded by the rules of our societies form true limits to our world just like those yielded by natural laws. True, not all the rules of our societies are internalized by all their members to the same degree; some of us have been brought up to simply ignore some rules or look at them as an inevitable evil. And also we may be able to *sometimes* assume a reflective attitude to the rules, in which we *do* see them as essentially different from natural laws, something that may not only be violated, but also questioned and possibly discarded. (This is, after all, what made Kant conclude that we have not only *Verstand*, but also *Vernunft*.) Each of us is nevertheless a social being in that we live in a world largely delimited by "soft," normative boundaries, rather than "hard," physical ones.

Returning to our earlier mentioned paleolithic hunter and contemporary clerk: the difference between their predicaments is highlighted if we imagine that the hunter was able to find food, but some tangible barrier lay between him and the food, preventing him from getting it. For both persons, the food is almost within grasp, but its grasping is obstructed. In the case of the hunter the obstruction is physical, whereas in the case of the clerk it is merely virtual, *normative*.

An objection might be that the second case does not truly differ from the first—that the physical possibility of taking the food from the supermarket shelf is only illusory, for it would lead merely to the supermarket's security taking the food away from the stealer (with other unpleasant consequences). But the fact is that it is often relatively easy and without much danger of recrimination to steal something in a supermarket—and despite this the clerk would probably still not do it, for he *does not want to steal*, i.e., he respects the norms of our society, according to which one *should not* steal.

Hence in both cases the protagonists are restrained by barriers provided by their environment—however, while in the first case it is a *physical* barrier straightforwardly yielded by causal laws governing the world, in the latter it is a *normative* barrier yielded by the fact that the clerk is bound with various kinds of rules. This indicates that besides the physical world, which restricts us in various ways, we also live in a kind of the virtual world (or worlds), which is superimposed over the physical one and yields surplus restrictions.

True, the normative boundaries are different from the physical ones in that they are "softer"—if our bank clerk really was *dying* of hunger, he would most probably break through the normative barrier and take some food from the supermarket shelf. (And given his condition, he might not even be called to account for this.) However, it is also true that the clerk, throughout his life, may never be in this kind of perilous situation—he will, more probably, face the normative barriers without the emergency license to break through them, and hence they will remain genuine barriers for him.

The continuation of the objection posed above might now be that the so-called normative barriers are only metaphoric descriptions for complicated causal mechanisms. Again, there is undoubtedly a sense in which this is true; but this sense is trivial. Of course, norms exist only in so far as people endorse them, and the endorsement of a norm can be tracked down to some patterns of activation in their brains. But there is no hope that we could exactly describe what is going on at a purely physical level. And, moreover, even if we could, it would hardly diminish the difference between the first and the second case: in the former one we have a straightforward physical impossibility of fulfilling one's desires, in the latter it is a vastly complex conspiracy of features of the physical environment (crucially involving the interaction of myriads of neurons in many human brains) yielding a much more complicated mechanism diverting the person from fulfilling it.

5 Conclusion

Humans' radical difference from their animal cousins can be seen as a matter of their having become truly uncanny niche constructors. Unlike other species, humans have not only tampered with their environment, but entirely rebuilt it. Moreover, along with the reshuffling of the existing elements of the environment and erecting brand new, artificial elements atop of them, "virtual environments" have been brought into being, in which the physical barriers of the natural world are largely replaced or supplemented by artificial ones.

What the artificial barriers of our "virtual worlds" are made of are *rules* (understood as social institutions that can affect people as powerfully as physical restraints). However, as I argued elsewhere (Peregrin 2010), rules are also at the core of the human ability to pass on behavioral patterns (not to mention "thoughts," "ideas" etc.) and thus, as Kant put it, to live not only within the realm of nature, but also within our distinctively human realm of freedom. In this way, rules appear to be the true key to human distinctiveness—their emergence, i.e., the emergence of our ancestors' abilities to think in the "normative mode," to acknowledge "ought-to-be's," has endowed us with a complex package of distinctively human features, including the ability to develop culture and to build virtual niches.

References

Boyd, R. and Richerson, P. J. (2008). Gene-Culture Coevolution and the Evolution of Social Institutions. In *Better than Conscious? Decision Making, the Human Mind, and Implications for Institutions*. C. Engel and W. Singer (eds.). Cambridge, Mass.: The MIT Press, 305-324.

Cavalli-Sforza, L. L. and Feldman, M. W. (1983). Cultural versus Genetic Adaptation. *Proceedings of the National Academy of Sciences of the USA* 79: 1331-1335.

Clark, A. (1997). *Being There Putting Brain, Body, and World Together Again*. Cambridge, Mass.: The MIT Press.

_____ (2008). *Supersizing the Mind*. Oxford: Oxford University Press.

Durham, W. H. (1991). *Coevolution: Genes, Culture and Human Diversity*. Stanford: Stanford University Press.

Feldman, M. and Laland, K. (1996). Gene-Culture Coevolutionary Theory. *Trends in Ecology and Evolution* 2: 453-457.

Flinn, M. V., Geary, D. C. and Ward, C. V. (2005). Ecological Dominance, Social Competition, and Coalitionary Arms Races: Why Humans Evolved Extraordinary Intelligence. *Evolution and Human Behavior* 26: 10-46.

Hutchins, E. and Hazlehurst, D. (1992). Learning in the Cultural Process. In *Artificial Life II*. C. Langton, C. Taylor, D. Farmer and S. Rasmussen (eds.). Reading: Addison-Wesley, 689-706.

Laland, K. N., Odling-Smee, J. and Feldman, M. W. (2000). Niche Construction, Biological Evolution, and Cultural Change. *Behavioral and Brain Sciences* 23: 131-175.

Nowak, M. A. (2006). Five Rules for the Evolution of Cooperation. *Science* 314: 1560-3.

Odling-Smee, F. J. (1996). Niche construction, genetic evolution and cultural change. *Behavioural Processes* 35: 195-205.

Odling-Smee, F. J., Laland, K. L. and Feldman, M. W. (2003). *Niche Construction*. Princeton: Princeton University Press.

Odling-Smee, J. and Laland, K. N. (2009). Cultural Niche construction: Evolution's Cradle of Language. In *The Prehistory Of Language*. R. Botha and C. Knight (eds.). Oxford University Press, Oxford.

Peregrin, J. (2010). The Enigma of Rules. *International Journal of Philosophical Studies*, 18: 377-394.

Richerson, P. J. & Boyd, R. (1998). The Evolution of Human Ultra-Sociality. I. Eibl-Eibisfeldt and F. Salter (eds.). *Ideology, Warfare, and Indoctrinability*. Oxford: Berghan Books, 71-95.

West, S. A., A. S. Griffin and A. Gardner (2007). Evolutionary Explanations for Cooperation. *Current Biology* 17: R661-R672.

15

Searle and Kim on Emergentism

Ranjan K. Panda

1 Introduction

Jaegwon Kim in one of his papers, "Emergence: Core Ideas and Issues" (Kim 2006), puts forth the argument of *downward causation* along with the *supervenience* and *irreducibility* thesis. For Kim, *downward causation* facilitates the demonstration of the supervenient relationship between the mental and the physical. The supervenience thesis along the line of emergence redefines the idea of overdetermination. Searle has been a great critic of supervenience and emergence theory of mind. The Searlean account of *mental causation* does refer to building up a link between the mental states and the brain states demonstrating the *downward movement*. However, this link is further intentionalized with irreducible mental features like intentionality, subjectivity, freedom, etc. His irreducibility thesis maintains the notion of emergence of mental states but does not find the notion of *supervenience* as a sound theorization of the mind. The chapter discusses the Searle's response to Kim's ideas about emergentism, critiques the Kimian viewpoint of the causal efficacy of the *bottom up* notion of mental causation and argues for a comprehensive irreducibility thesis, highlighting the significance of biological naturalism.

The chapter is divided into four sections. In the first section I begin with the introduction of the notion of emergence. And following that, in the second section, I bring out Kim's view on some of the key points of the notion of emergence with reference to mind-body/mind-brain relationship. The Searlean notion of emergence and supervenience is discussed in the third section. The last section brings out an overall assessment of the Kimian contention of emergence and the merits of Searlean biological naturalism.

2 The Concept of Emergence: A Debate

The concept of emergence has a long history in philosophy in general and philosophy of mind in particular. This discourse flourished especially in 1920s when many philosophers like C. D. Broad, C. Lloyd Morgan, A. O. Lovejoy and many others tried to apply the notion of emergence to the problem of Cartesian dualism. The Cartesian dualism, as we all know, demarcates the mind from the body, understanding them as categorically different in their nature and therefore classifiable as two different substances. Yet the two substances with no mode of relating to one another create a problem. In recent time, Kim, Searle and many others, who have been advocating naturalistic worldview in philosophy of mind, are trying to deconstruct the Cartesian worldview. They want to treat the mind as a natural phenomenon emerging in the natural world. In other words, the mental is caused by the physical. Naturalism provides a scientific understanding of reality, which includes both the physical and the mental. Naturalism as the theory of mind is grounded in certain principles of the physicalist theory, which systematically unifies the mental states and processes with the physical states and processes. This unification is based upon the scientific methods of natural sciences to confirm objectivity, repudiating the truth claim of subjective or intuitive experiences (see Clark 1990, 157). Therefore, within the naturalistic framework the understanding of the mental goes along with the understanding of the physical. The physical composition and configuration of the body provides an ontological ground to investigate the nature of mental phenomena. The former has drawn the attention of many. For naturalists the mental is ontologically dependent on the physical and can be investigated by investigating the neurophysiology and the neurochemistry of the brain. Like any other material body, the brain is also constituted by certain physical properties. For naturalists, the function of neural processes causes the mind.

The notion of emergence treats the relationship between the brain and the mind as that of cause and effect, advocated on an implicit background of evolution. The evolutionary process demonstrates various levels of this relationship, starting from the material existence of atoms and molecules to the emergence of organic body and then to the occurrence of the embodied beings. The human as an embodied being not only shows the body as biologically evolved, but also demonstrates the emergence of the mind. In other words, human mind is treated as the *emerged* property of the neurophysiological system of the human body. The emergence of the body and the mind also further indicates a functional relationship that exists between them. Philosophers of mind and cognitive scientists try to interpret the nature of the human mind or consciousness by looking at these functional relations. It is a matter of some debate whether the causal relation between the mental level and the physical level is a one-way process or whether it operates in both ways. According to Kim, the causal relation between the physical and the mental is a one-way process only. On the other hand, Searle advocates that the causal processes operate on both levels of functions. Moreover, the physical causal relations and the mental causal relations are operating at two different levels. Since the mental is caused by the physical conditions of the bodily system, the question that comes to the minds of some philosophers, is that whether the emerged mental phenomena can be reduced to the physical level of functions. Kim and Searle hold different viewpoints in this regard. Kim argues consistently for supervenience and irreducibility of the mental. Searle also holds the notion of irreducibility of the mental phenomena, but he points out that the supervenience thesis of Kim characterizes the mind as an epiphenomenon. In this regard, along with the functional causal relationship, the notion of irreducibility and supervenience become debatable issues in the discourse of the emergence theory of mind. To fully understand the Searlean critique of emergentism, one has to take notions, such as mental causation, irreducibility and supervenience, as all roped together.

One may concede that human mind is an emergent phenomenon. That is, it has emerged in the natural processes of evolution and its functions are rooted in the functions of the neurophysiology of our brains. In that respect, the mind is conceived in continuum with the evolving conditions of the brain and the environment. For an emergentist, the mind is conceived as a whole, constituted by various functions of the parts or the subsystems of the brain. Nevertheless, the emergence of the *whole* is *qualitatively* different from its constituted *parts*. The organization of "the parts into wholes may be called creative synthesis."[1] As Baylis briefly describes it, "Emergents are those characters of a complex which are not also characters of a proper part of that complex, and emergence or creative synthesis is that event which occurs when the complex having emergent characters is formed" (Baylis 1929, 373). In other words, the emerged properties are not only different from the complex system features but also can constitute a complex system. This difference is due to the effect of *emergence* followed by *submergence*. For example, the emergence of water is caused by the integration of two molecules of hydrogen and one molecule of oxygen. In the process of integration of these molecules, the molecules lose the character of being a gas. The process of losing the old character is called *submergence*, whereas the process of integration of development of a new character is called *emergence*. Emergence and submergence result from integration and disintegration (ibid., 374-375). Since we refer to the complex notion of emergence, it implies that multiple numbers of entities and their relations could integrate for causing a complex phenomenon, and in the process of integration they also destroy other properties. Therefore, integration and disintegration of elements for each such a change means breaking of old relations and forming of new ones (ibid., 376). This could be accepted as a reason for a differential gap between the cause of

---

[1] Cf. Baylis 1929, 373. Baylis in this connection refers to E. G. Spaulding's notion of "creative synthesis."

emergence and its new effects. In other words, the consolidated function of the emergent properties can further constitute a complex system function, revealing that emergent properties are causally efficacious to form new relations.

Assuming that the emergence is a continual process implies a development of successive levels and their relationships. The levels of relationship and the comprehension of a novelty embedded in the structural configurations are attributed to the molecular level of existence of the phenomenon. The levels of existence of beings and their functions are manifested in the physical world. As Ablowitz writes, "The application of this to the physical world results in the hierarchy of atom–molecules–crystal. Just as the combination of atoms into a molecule reveals new and unpredictable qualities, just so does the structural organization of molecules into a crystal or colloid reveal characteristics which are non-significant or impossible at the molecular level, but attributable to the crystalline or colloidal level" (Ablowitz 1939, 2). Crystals, molecules and atoms are the micro level constituents of the material system. The organic body would also have different layers or levels including the micro levels. The hierarchy of levels becomes more complex when we talk about the human being as an embodied being. The micro and macro levels broadly refer to the level of the brain and to the level of the mind respectively. The brain functions at the micro level causes the mental level functions, such as the formation of representations, beliefs, desires, intentions, etc. The complexity involved at each level of functions shows a sort of *leap* from one level of emergence to another. These leaps indicate *transformation* (ibid., 4) undergone in the emergent properties at different phases of the process of emergence.

One of the pertinent issues, in this regard, is the notion of *predictability*. The prediction of one level of function is done by looking at the emergent conditions of its base level patterns. For many brain scientists, it is a *temporal* phenomenon as they are rigorously investigating the patterns of brain functions. They are of an opinion that over a period of time, the mystery of the mind can be explained away.[2] The explanation of the mental calls in the explanation of the physical. Explanation and prediction coincide with each other when we delve into the study of underlying properties of the brain and their respective functions. The novelty involved in the properties of emergence can be functionally integrated with the underlying causal properties. This integration is a matter of *mutual* relations required for the *explanation* of the new qualities or the emerged system features (Ablowitz 1939, 9). To understand the mutual relations we need to formulate laws at the organic levels of functions or little below the level at which organic activities are shown in terms of physico-chemical reactions in the brain. What one may take into account is "the degree of appli-

---

[2] One of them is Ramachandran (2003).

cation of the law." Nevertheless, such lawful explanation would be encompassing the explanation of any phenomena in the naturalistic worldview (ibid., 10).

3 Emergence, Supervenience and Epiphenomenon

Kim advocates the theory of emergence within the naturalistic paradigm. He makes three important interventions while interpreting the notion of emergence with a reference to the philosophy of mind. They are, firstly, points of reciprocity of emergence and supervenience; secondly, finding out the functional correlations of the emergent properties at basal level conditions; and thirdly, emergence and downward causation as complementary to one another for their respective explanations. Kim accepts that emergence is possible at a certain level of complexity of the physical system. The pattern of complexity gives birth to emergent properties. So far as the first idea is concerned, emergent properties supervene on the micro level or the *basal* level structural properties. Supervenience is a relation that emergent properties share with the basal level structure. As Kim defines supervenience, "If property M emerges from properties $N^1 \cdots N^n$, then M supervenes on $N^1 \ldots, N^n$. That is to say, systems that are alike in respect of basal conditions, $N^1 \ldots, N^n$ must be alike in respect of their emergent properties" (Kim 2006, 550). In other words, only the human brain functions can cause the human conscious mental states that emerge *from* certain firm condition of the neural substrates function of the brain. The stability in function shows that emergent properties are not randomly caused; they rather have "deterministic condition" in which their activities could be statistically measured. Let us assume that the feeling of pain is the simulation of c-fibre. This simulation is identified with regular activities at the neural substrate of the c-fibre. The regularity condition gives a determinative idea about the basal level function of the neural substrate that causes the mental phenomena. Moreover, the notion of supervenience recommends "asymmetric dependence of the mental and the physical" (Kim 2000, 6). This dependency of the supervenience relationship of the mental is defined by the causal asymmetric relations. Furthermore, the dependency is *indiscernible*. The mental being supervenient on the physical implies that the physical has the mental; more specifically, the idea is that the mental is realized in the physical.

Kim makes his second intervention at this point, emphasizing determinating conditions of emergence and referring to the mental states and their neural correlations. The detection of the neural correlation-function would help in *prediction* and *explanation* of emergent mental states. For Kim, "prediction and explanation is a matter of deduction or derivation" (Kim 2006, 550). One can draw a definition looking at the *functionalization* of the *properties*. The functional definition covers three things; in the case of pain, these are pain, the brain and the behaviour. The *feeling* of pain is a mental state causing pain behaviour. Between pain behaviour and *feeling* of pain one needs to show „how does one *realize* pain?" So, the detection of the *neural realizer* is an essential aspect of the explanation of pain. Pain, like any other emergent mental phenomena is *realized in* the neural states of the brain. The realization process functionally relates to the pain and pain behaviour. Kim argues for the "primacy of the physical and its laws without implying physical reductionism, thereby protecting the autonomy of the mental" (Kim 2000, 11). Laws are construed as derivations of a functional definition in order to locate the relationship between emergent properties and their neural correlations. The derivation weakens the thesis of irreducibility as a condition of the emergent mental properties. It rather suggests that the derivative law can be formed with the help of tracing out the neural structure and function at the base level.[3]

The micro level functions are importantly characterized in two modalities, they are, the mode of emergence and the mode of realization. In the mode of emergence the physical basal conditions do not establish a *homogeneous* natural relationship with the mental properties. In other words, there is no natural dependency that occurs in the case of emergent phenomena and their causal conditions. The supervenience dependency thus entails that the mind is an epiphenomenon without being causally empowered. The causal efficacy of the basal level physical conditions of the brain, however, remains a brute fact to explain the notion of emergence. Kim's third intervention at this point not only intends to show that the downward causation is indeed a non-efficacious causal relation, but is required for fulfilling the conditions of functionalizability. That would explicate whether supervenient properties could cause a downward relation in order to establish a causal chain between the physical level cause and physical level behaviour. In other words, let's assume that if the physical cause is p and the physical level behaviour is p^*. Can we then say that emergent property M is *caused by* p and thereby *causing* the p^* Kim rules this out. He writes, "Moreover, it is not possible to view the situation as involving a causal chain from P to P^* with M as an intermediate causal link. The reason is that the emergence relation from P to M cannot properly be viewed as causal."[4]

---

[3] "Conscious experiences, or anything else for that matter, are reducible if and only if it is functionally reducible, and it is functionally reducible only if it is functionally definable or interpretable." See, Kim (2006, 553-554).

[4] Ibid., 558. Giving a footnote to the notion of causal relation, Kim refers to Morgan and writes, "Morgan explicitly denies that emergence is a form of causation (Morgan 1923, 28). Moreover, there is little to recommend in the claim that a neural state causes pain and then pain in turn causes, say my hand withdrawal. How can there be a causal chain from pain to the hand motion that is separate and independent from the physical causal chain from the neural state to the motion of the hand?"

Kim does consider that mind is an emergent property of the physical, but this emergence does not show the causal link between the two levels of function. The mental experience supervenes on the physical, and it is has *emerged* rather than been *caused by* the neural relations. He further writes, "This appears to make the emergent property M otiose and dispensable as a cause of P^*; it seems that we can explain the occurrence of P^* simply in terms of P, without invoking M at all" (ibid., 558). The existence of the mental is *excluded* so far as the explanation of behaviours or actions are concerned, but *included* as an *epiphenomenon* having *supervenient* and *indiscernible* relationship with the world. This inclusion is under the coverage of a *strong closure of causal relations,* which only incorporates the events in the natural world.

4 From Emergence to Irreducibility: a Reading of Searle

John Searle also develops the idea of *causal closure* while advocating *biological naturalism.* The biological naturalism as a theory of mind maintains the view of emergentism. The mind has evolved biologically along with the human body constituting of atoms and molecules. This biological creation is a part of the natural world. The human body is governed by the natural laws. The natural laws draw the *causal closure,* unfolding the naturalistic project of Searle's theorization of the mind. Moreover, human as an embodied being requires special explanation with regard to the emergence of the mental phenomena. In fact, there are two significant aspects to the Searlean theorization of the mind conceived within the naturalistic paradigm. They are the notion of *emergence* and the notion of *irreducibility*. For Searle, consciousness or the mental phenomena like intentionality, subjective feeling, freedom, etc., are the emerging properties. As emergent properties they have *appeared* out of the causal function of the brain processes. The causal functions sustains biologically between the brain processes and the mental processes. Searle hypothesizes this causal process in his *biological naturalism*. The hypothesis advocates, "Mental states are *caused by* the brain processes and *realized in* the brain processes" (Searle, 1986, 262-272). The brain is the locus of causing and realizing the mental states. In other words, processes of *causing* mental states and *realizing* them are happening at the realm of the physical. In that domain mental states have their own intrinsic features like intentionality, phenomenality, subjectivity, etc. These mental features remain distinct and irreducible to the mere neurophysiology and neurochemistry of the brain. The brain being the constitutive patterns of these layers of functions, cause the emergence of the mental and prevails on *causing* the behaviour. Searle illustrates, this causal emergence of the mental as follows:

> Consciousness is causally emergent property of systems. It is an emergent feature of certain system of neurons in the same way solidity and liquidity are emergent features of system molecules. The existence of consciousness can be explained by the causal interaction between elements of the brain at the micro level, but consciousness cannot itself be deduced or calculated from the sheer physical structure of the neurons without some additional account of causal relations between them.[5]

The brain being the micro level constituent of consciousness is different from the emergent mental states. The mental states are conscious and belong to the macro-level. This micro-macro stratification requires different types of explanation though they functionally operate in the realm of the physical. As the macro level function does affect the behaviours of the person, the micro level does cause them and is involved in realizing them.

The involvement of the physical is inevitable as a constituting cause of the mental. But this constituting cause does not fully suffice to explain the intrinsic features of the mental states and their interaction. Explanation of their interaction requires an additional account of causal relations. Moreover, that may not entirely put off Searle's thesis of irreducibility. For him,

> consciousness can be caused out of the brain processes but it would have certain „causal power that cannot be explained by"[6] the micro level system features of the brain. This thesis would entail another notion of emergence, that is, "... consciousness could cause things that could not be explained by the causal behaviour of the neurons." The naïve idea here is the consciousness gets *squirted out* by the behaviour of the neurons in the brain, but once it has been squirted out, it then has a life of its own. (Ibid.)

Consciousness, in this respect, as an emerged system feature, is causally efficacious to build a system and determine the function of the mental functions entirely. In other words, this notion of emergence would imply the autonomy of the mental that could be maintained out of the functions of the brain processes. Searle's naturalistic schema does not approve it.

However, he intends to argue that the notion of irreducibility of the mental stating that

> Consciousness is causally emergent property of the behaviour of neurons and so consciousness is causally reducible to the brain processes. But—and this is what seems shocking—a perfect science of the brain would still not lead to an ontological reduction of consciousness in the way our present science can reduce heat, solidity, colour and sound. (ibid., 116)

---

[5] This notion is Emergent 1, Searle (1992, 112).
[6] This notion is Emergent 2. Ibid.

Searle lays emphasis on non-reductionistic theory of mind showing that reductionism ultimately advocates ontological reduction. The ontological reduction of the mind to the brain processes eliminates the reality of the mental phenomena or consciousness. The reality of consciousness lies in the first person subjective experiences and feelings. This reality appears in the macro level. In other words, the *appearance* of consciousness in our feelings, experiences and actions is not only different from the physico-chemical processes of the neurons of the brain but also shows that the appearance is reality (ibid., 121).

However, the sustained causal interaction between the brain and the mind shows the upward and the downward movement of the causal functions respectively. As the causal power of the brain is causing consciousness, similarly the *causal power of the mind* is responsible for voluntary movement/action, feelings, experiences, etc. The causal power of the mind shows a downward mobility causing/*transforming* mental states into actions. The existence of the mental hence supervenes on the physical states and processes of the brain. Searle writes,

> On this characterization of the supervenience relation, the supervenience of the mental over the physical is marked by the fact that physical states are causally necessary for the corresponding mental states.... this sort of supervenience is a causal supervenience (ibid., 125).

The causal supervenience claims that the mental interacts with the physical.

5 Critiquing Kim's Notion of Emergence

Interestingly, Kim denies the causal interaction between the mental and the physical. The mental, being supervenient on the physical, cannot have causal impact on it. Rather, mental properties are *causally impotent* and do not imply reducibility. It is because there is no causal link between supervening features of the mental and its basal level conditions. The supervenient relation only expresses a *vertical* dependency relation. But this expression has nothing to do with the real *horizontal* causal relations that the base level holds while causing another mental/brain state (e.g. *P causing P**). While arguing for supervenience, Kim understands that "it is not possible to have causation in the mental realm without causation being crossing into the physical realm." Its significance lies in a state of relation where „we can no longer isolate the causal relation at higher level and its bottom level which would entail cross level causal relation." Neither Kim would talk about the cross level causal relationship between the physical and the mental nor would he talk about the level bound autonomy, which would be "inconsistent with the notion of supervenience" (Kim 2003, 156). If his notion of supervenience is acceptable then it does not fit with the notion of emergence. The notion of emergence shows the *gap* between two levels of complex functions. Each level function will have causal impact, i.e. the power of making a complex structure and further giving rise to certain emergent properties. This process may not have direct causal link but one getting emerged out of the other and being able to sustain a complex function itself causally empowers the emergent properties. This is very much a part of the Searlean theorization of intentionality emerging as intentional mental feature out of the functions of the brain. Intentionality of mental states builds up the network of mental states just being empowered with certain causal power. This intentionality is not thrown out of its material ontological basis. The gap between the causal process of the physical level and the causal process of the intentional mental level is a conceptual gap, which any emergence theory would advocate. Kim in this regard is an exception to this idea. In Ned Block's terminology, the causal power of the emerged mental property *drains away* (Block, 2003) making the mental epiphenomenal.

On the other hand, if supervenient causal relationship rests on the horizontal position then there is no *level* of emergence *per se*. Assume that one level of emergence is the level of the mental; it would only constitute a parallel mental level as a supervenient level. At this level, we would have the emergence of discrete mental states corresponding to change and occurrence of the brain states at physical basal level. Emerging mental states would be discrete in the sense that they would not have any *unifying* properties to relate with. Even if we maintain the condition of simultaneity and multiple realization conditions for emergent states and their realization, lack of unifying character to build them would still imply the loss of *subjectivity*. Therefore, the causal chain that builds in is only at the level of physical which excludes the mental. Kim writes

> Note that P's causation of P^* cannot be thought of as a causal chain with M as an intermediate causal link; one reason is that the P - to - M relation is not a causal relation. Note also since that M supervenes on P, M and P occur at the same time. (Kim 2003, 157)

This notion of *exclusion* of the mental not only eliminates the *feeling* of *subjectivity* but also eliminates the notion of *level per se*. That is, drawing up differences in the levels with horizontal causal mobility may imply emergence but would not imply levels. How does one talk about vertical level of emergence with reference to the formation of water? The two levels of existence are causally efficacious and functional so far as their *emergence* and *submergence* are concerned.

The notion of *simultaneity* needs to be discussed with regard to the notion of *overdetermination*. Kim does not want to defend the notion of *overdetermination* because it involves the causal powers of the mental states along with the physical state for determining conditions of *realization*. Mental properties are *extrinsic* and *relational*. They are extrinsic because "no mental property would have non-physical realizations. Rather they must be physically realized" (Kim 2000, 19).

Mental properties are "second-order functional properties" relative to the "first-order physical functional properties" (ibid., 21). The former do not constitute the functional properties; rather they share nomological relation with the latter. If mental properties intervene in relationship with the physical for determining condition of realization, then it implies that the physical supervenes on the mental. There is no such supervenience justified because it would entail reductionism. Thus, for Kim, overdetermination entails reductionism. He finds that the Searlean mental realism advocates overdetermination. It is because the mind intentionally or *consciously* acts *in* the world (Searle 1992, 107). The intentional mode is effectively felt and shows the realization of the mental states. The notion of overdetermination with reference to *experience* and *realization*, shows *fusion* of top-down and bottom-up causation. This fusion represents the notion of simultaneity: Say, "I am feeling pain." It means I have *sensory qualia* of pain, which simultaneously *go* with the *intentional qualia*.[7] The latter implies a *dynamic* intentional representation of *feeling* of pain, whereas the former represents a *stable* sensory representation of pain. For Searle, both are two different modalities of processing the representation. Moreover, the feeling of pain needs intentional representation to get relieved from it. Then "how do I make this representation?" is certainly an important starting point of referring to thoughts and their dynamic relations, which helps in forming a synthesis/unity of those representational states. It is obvious that the process of synthesizing would have physical basis. The time involved in the physical simulation (whatever milliseconds) is unlikely to be at the same time involved in intentional representation. The simultaneity *reduces* the differences in the processes. If simultaneity zeroes down the levels of processing and levels of interferences between the intentional and the sensory activities, then Kim misses out the "homogeneous natural relations" between the physical and the mental. In this regard, supervenience would entail reductionism.

The Searlean thesis does not imply overdetermination, but rather a functional correlation, thereby implying the functional autonomy. The notion of a functional correlation between the top-down mental causation relating to the bottom-up causal processes is a conjecture. We are yet to *know* their level of meeting points, and Searle candidly admits that. Where do the physical processes end with emergent mental properties and where do the mental properties reside with the physical, is not only a matter of scientific investigation, but also an issue to be delved through philosophical inquiry. Even if we empirically find their point of emergence and point of correlations, still the significance of

---

[7] Ken Mogi's lecture on "Qualia and Contingency," International Workshop on Models of Brain and Mind: Physical, Computational and Psychological Approaches, held from 21–24 November 2006, at Saha Institute of Nuclear Physics, Kolkata.

the mental will not be nullified; it would have its own level of description. The Searlean notion of levels of description is therefore significant, but does it not divide reality, falling into the trap of property dualism?

However, the notion of property dualism could not be strengthened due to Searle's disclaimer for *ontological emergence* (see Silberstein and McGeever 1999). Searle, like Kim accepts the "modest kind of emergence."[8] Advocating the modest notion of emergentism, he may succeed in showing that ontology of intentionality or the mental could be traced from the ontology of the physical. Moreover, in Searle's interpretation, the mental has the power (causal power) to go beyond the *causal closure* held by the physicalists. Searle is neither a physicalist, nor is he a radical naturalist. His naturalism does not advocate reductionism under the banner of unification of science. Nevertheless, he does not disown the laws of the nature operating in the universe and having impact on the body. But, at the same time, we are involved in performing conscious activities, relating ourselves with the world, forming norms and defining various types of rationalities involved in human social life. All these have a major epistemic significance and are formed by the "conscious intentionality." The notion of *conscious intentionality* becomes a ground for experiencing causal relations in the world. The emergence of intentionality as an intrinsic mental feature fulfils the epistemic necessity of beings like us. Our epistemic needs are satisfied by the emergent mental properties. The Searlean thesis argues for a higher order explanation of these activities. Then, is it the case that Searle is advocating "epistemological emergence" (Silberstein and McGeever 1999, 182)?

Defining the notion of epistemic emergence Silberstein and McGreever write:

> A property of an object or system is epistemically emergent if the property is reducible to or determined by the intrinsic properties of ultimate constituents of the object or systems, while at the same time it is very difficult for us to explain, predict, derive the property on the basis of ultimate constituents. Epistemologically, emergent properties are novel only at the level of description. (Silberstein and McGreever 1999, 186)

In other words, the epistemic emergence talks ultimately about the reduction of the emerged properties to its basic constituent properties of the system. Following this definition of epistemic emergence, Searle's notion of intentionality, subjectivity, etc. as emerged system features/properties are emergent properties and hence ultimately reducible to the function of the brain. Though Searle is hopeful about a scientific explanation of consciousness,[9] and determination of

---

[8] See Kim's classification of the notion of emergence in his (2006, 549).

[9] In Searle words, "I believe with recent progress in neurobiology and with a philosophical critique of the traditional categories of the mental and the physical, we are getting closer to being able to find a systematic scientific way to answer this question [what is the relationship between

its relationship with the brain, still he is not in favour of any kind of reductionism. According to him, human intentionality and subjectivity are irreducible features of consciousness. As Searle primarily puts it

> Consciousness is, by definition, subjective, in the sense that for a conscious state to exist it has to be experienced by some conscious subject. Consciousness in this sense has a first-person ontology in that it only exists from a point of view of human or animal subject, an "I", who has the conscious experience. Science is not used to dealing with phenomena that have first person ontology. By tradition, science deals with phenomena that are "objective," and avoids anything that is "subjective." (Searle 1999, 2074)

Hence, for Searle the relationship between subjectivity, intentionality and consciousness needs to be explained from a different macro level, whereas the relationship between the consciousness and the brain is to be viewed from the micro level.

Now the question may arise, are these two levels of functions independent of one another or are they interrelated? These functional explanations are independent of each other because they explain things from two view points: the macro level explanation of consciousness or conscious action is from the *first person point of view*, whereas the micro level explanation of human action is from *third person point of view*. The epistemic objectivity of these explanations differs because the first explanation presupposes human subjectivity as an objective fact of the reality. The epistemic objectivity stated from third person point of view shows there are entities which have objective existence. These entities do not depend on someone's subjective opinion or point of view. Rather, they have an independent existence in the world. So, these points of view bring in the "distinction in modes of existence, i.e. the ontological sense of the objective subjective distinction" (ibid.). Searle makes it clear that this epistemic objectivity of science does not include the ontological subjectivity. In this regard, the notion of independence can be discussed with reference to two different frameworks in which the emergent conscious features are viewed. They are the framework of science and the framework of philosophy. Moreover, if the explanations of these frameworks are interrelated, then Searle is offering a functionalist explanation of the mental phenomena. But, obviously, Searle does not accept functionalism. Though mental phenomena are caused by neural configurations and realized in the same neurological structure, there is no identity as such between the neurophysiological (micro) level and the mental (macro) level. Thus the mental is not just an emergent property, but is something real.

---

consciousness and the brain process?]" See (Searle, 1999).

Furthermore, this reality of consciousness could presuppose the ontological emergence in which the mental features likes subjectivity, intentionality, qualia, etc. are having a unified ontological status. So far as the notion of ontological emergence is concerned, Silberstein and McGreever state that

> features are neither reducible to nor determined by more basic features. Ontological emergent features are basic features of systems or wholes that possess causal capacities not reducible to any of the intrinsic causal capacities of the parts, nor to any of the (reducible) relation between the parts. (Silberstein and McGreever 1999, 186)

Looking at this notion of irreducibility thesis prevailing as one of the strong points in the framework of Searlean theorization of emegentism, David Papineau wondered during my talk in Prague in 2009, whether I was not assuming a notion of metaphysical supervenience. According to Papineau, metaphysical supervenience presupposes the mental as an ontological category, prior to the ontology of the physical. The answer to such a position ultimately leads to the traditional philosophical position of the Cartesian dualism. Searle is a great critique of the Cartesian metaphysics. He writes:

> As long as we continue to talk and think as if the mental and the physical were separate metaphysical realms, the relation of brain to consciousness will forever seem mysterious, and we will not have a satisfactory explanation of the relation of neuron firings to consciousness. (Searle 1999, 2074)

For him, all the evolutes are causally connected to the nature as a whole. The mind is seen as part of the nature. But this relationship does not hold to any sort of reducibility, that mind or the conscious experience that comes in the unified form is casually reducible to the physical structural properties and functions. It follows that there is a relationship between the mental and the physical but one can never be reduced to another. They belong to the realm of the biological, representing two different layers of existence.

The human body is a complex biological organism. So far as the mind is concerned, it has evolved out of the complexity involved in the evolutionary process in general and the neurophysiolgoical structure of the brain/body in particular. The emergence of mind from the neural structure makes Searle to hold the view that mind is dependent on the brain (the most complex part of the body), where the mental phenomena are caused by the brain. But there is no causal law operating in the realm of the mental. On the other hand, actions are intentionally defined. Intentionality is one of the basic features of the conscious mind. Thus, the relationship between the mental and the physical is one of the asymmetrical dependencies.[10] It shows that the mental is dependent on

---

[10] This notion of dependency, according to John Heil, is also known as the *metaphysical depen-*

the physical but not reduced to the physical. In this regard, Searle's philosophical position comes close to Kim's nonreductive physicalism.

However, one needs to go back to the unified feature of human consciousness or human intentionality *per se* to grasp the *nonreductive* standpoint of the Searlean thesis. Experiences manifest unified intentionally which makes a space for not only asserting the ontological but also postulating the notion of the self (see Searle 2004, 201). As we have stated earlier, for Searle, "all conscious states are ontologically subjective because they have to be experienced by a human on an animal subject in order to exist" (Searle 1998, 1937). Subjectivity is intrinsic to experience and presupposed as an objective phenomenon. The disclosure of subjectivity in experience, thought and action shows the postulation of the self as an intentional being. This self is not only a thinking being but also a performer of free action, rational choice, decision, etc. Thus, the self, as the representation of intentional subjectivity, involves the activities of thinking as well as action in the form of an interaction with the world. The intentional subjectivity is constituted of person's attitudes, feelings, experiences, moods, etc. Human thinking is not only intentional, but also normative. Being intentional it demonstrates how the content of thought and expression is experienced and understood.

The normativity of intentionality shows how human beings form institutions and live a value laden life in society. Searle calls it the *deontic power* (Searle 2007, 125) of human intentionality. It refers to the normative ability of judging the content of thoughts and action. This is something significant to human beings engaged in knowledge creation. This ability pertains to all kinds of knowledge such as scientific, cultural, religious, moral, etc.–it is the power to evaluate the very content of knowledge and the framework in which knowledge is cultivated. The ability to exercise the power of judging the right and wrong; the will to perform certain duties; desire to show empathy, aspiration to know the truth, etc., unfold an important mode of the function of human intentionality at the mental and the social level.

Studying the notion of emergence one finds an interesting nexus that operates in various levels, starting from the biological to the mental and then to the social level that unfolds mystery of consciousness embedded in the complex structure of life. The metaphysics of evolution not only reflects the emergence of these levels in the evolutionary process but also shows the relationship of consciousness with the phenomenon of life. The above discussion on emergence shows the mind is not only affected in the evolutionary process but also is a *nonreductive* phenomenon in the history of evolution.[11]

---

dency. See, Heil (1998, 42).
[11] I express my gratitude to Juraj Hvorecký and Tomáš Hříbek for giving me a wonderful op-

References

Ablowitz, R. (1939). The Theory of Emergence. *Philosophy of Science* 6 (1): 1-16.

Baylis, C.A. (1929). The Philosophic Functions of Emergence. *Philosophical Review* 38 (4): 372-384.

Block N. (2003). Do Causal Power Drain Away? *Philosophy and Phenomenological Research* 67 (1): 133-150.

Clark, P. (1990). Explanation in Physical Theory. *Philosophy*, Supplement 27: 155-175.

Heil J. (1998). *Philosophy of Mind*. London: Routledge.

Kim, J. (2000). *Mind in a Physical World*. Cambridge, Mass.: The MIT Press.

_____ (2003). Blocking Causal Drainage and other Maintenance Chores with Mental Causation. *Philosophy and Phenomenological Research* 62 (1): 151-176.

_____ (2006). Emergence: Core Ideas and Issues. *Synthese* 151: 547-559.

Ramachandran, V. (2003). *The Emerging Mind*. London: Profile Books.

Searle, J.R. (1983). *Intentionality: An Essay in Philosophy of Mind*. Cambridge: Cambridge University Press.

_____ (1992). *The Rediscovery of the Mind*. Cambridge, Mass.: The MIT Press.

_____ (1998). How to Study Consciousness Scientifically. *Philosophical Transactions: Biological Sciences* 353 (1377): 1935-1942.

_____ (1999). The Future of Philosophy. *Philosophical Transactions: Biological Sciences* 354 (1392): 2069-2070.

_____ (2004). *Mind: A Brief Introduction*. New York: Oxford University Press.

_____ (2007). *Freedom and Neurobiology: Reflection on Free Will, Language, and Political Power*. New York: Columbia University Press.

Silberstein, M. and McGeever, J. (1999). The Search for Ontological Emergence. *The Philosophical Quarterly* 49 (195): 182-200.

---

portunity to present an early version of this chapter at the conference. And, many thanks to all the participants of the conference for their valuable questions and comments, and especially to Professor David Papineau, Professor Jonathan Knowles, and Professor Vikram S. Sirola for giving a deeper reflection to problems in my presentation and for their valuable suggestions.

16
Genealogy, Evolution, and Morality
Christoph Schuringa

Various arguments have been advanced that seek to draw conclusions about morality from facts about evolution. Sometimes they have sought to bolster morality, sometimes to discredit it. I shall try to show that these arguments fail to achieve the task they set themselves. In order to do so I will begin by considering the scope and limits of Nietzsche's genealogies. It will become clear that Nietzsche frames these genealogies in a specific way in order to avoid a fallacious form of inference. I go on to show that evolutionary arguments for or against morality are in danger of committing precisely this same fallacy. I conclude by assessing the prospects for arguments of this type. I do not anywhere intend to draw an analogy between Nietzschean genealogies and evolutionary accounts, or suggest that they involve essentially the same strategy applied in different domains (say, human history and biological history). I wish only to extract certain insights from a consideration of Nietzsche's genealogies and to apply them to the evolutionary case.[1]

The approach that I will adopt obviates the need to press the charge that has been most frequently applied to inferences from evolution to morality, that of illicitly deriving an "ought" from an "is," or committing the so-called "naturalistic fallacy."[2] Such a line of criticism arguably raises more problems than it solves. Firstly, it is by no means self-evident that inferences from factual premises to normative conclusions are always fallacious. Furthermore, it is not clear that arguments of the type under scrutiny would be touched by such a charge. It seems to me that they need not issue any moral imperatives or make normative claims in their conclusions; all they need to argue is that a certain

---

[1] There is a considerable literature on the issue of Nietzsche's complex relationship with Darwinism. Some authors, notably Richardson (2004), argue that Nietzsche's thought exhibits important Darwinist elements, despite his frequent criticisms of Darwin. This debate is too intricate to be entered into here.

[2] These charges are often treated as the same, although what the inventor of the term "naturalistic fallacy," G. E. Moore, meant by it seems to be highly distinct. The charge is also sometimes characterised as a breach of "Hume's Law" (although Hume probably never formulated such a law).

conception of morality is justified or unjustified. Be this as it may, however, I will show that the arguments in question fall at an earlier hurdle, before the status of statements involving "ought" even comes under consideration. It will be considerations about the nature of the "vindicatory" and "debunking" strategies, respectively, that will threaten these arguments, not considerations about the nature of claims about morality.

Accordingly, there will be little need to consider closely the nature of the "morality" these arguments are concerned with. This circumstance is helpful, given that determining what is meant by morality in these contexts is beset by a host of problems, exacerbated by a tendency in much of the literature to speak of "morality" in a vague or, what is worse, an ambiguous or shifting manner. At least three sources of confusion may be identified. First, some authors identify "morality" with a set of behaviours held to be morally laudable (typically, cooperative and/or self-sacrificing behaviour), while others take "morality" to involve the exercise of a capacity for moral reflection or deliberation which is supposed to underlie these behaviours. Secondly, some authors think of "morality" as a specific set of moral commands or principles (often equating them to something like a "common core" which all moralities must possess), while others leave the specific normative content of morality open. Thirdly, there is disagreement over the metaphysical status and "objectivity" of moral claims, and over whether metaethical commitments impact on how and whether the arguments can be made.

For present purposes, these issues do not need to be resolved. The discussion will be restricted to considering what evolutionary arguments can establish, and this will be found to fall short of claims about morality, whichever of the conceptions alluded to above may be in play. Rather, the emphasis will be on the nature of the evolutionary accounts themselves and the inferences that can legitimately be drawn from them. Here we avoid the dubious policy of presupposing some conception of morality and then showing morality to be an "illusion" because evolutionary explanations cannot account for it (cf., e.g., Ruse and Wilson 1986).

1 Genealogy

In his *On the Genealogy of Morality* (1887), Nietzsche was concerned to show how our moral valuations had come to be. The work announces itself in its subtitle as a "polemic," and thus as directed against some received opinion deemed worth attacking. One might think that it offers "a genealogy" that serves to mount an argument against this target. In recent discussion, genealogy has indeed often been conceived of as a type of argument, or even a special method, which has this task. This is mistaken, however.[3] A genealogy, for Nietzsche, is neither an argument form nor a method. Nietzsche clearly intended his genealogies to have critical force: they form part of the project of preparing for his "revaluation of all values." Genealogy does not, however, possess a "philosophical function" that allows it, by the force of its argument, to do its critical work.[4]

As Nietzsche repeatedly emphasises, the work of genealogy is intended as preparatory to a further project, that of the revaluation of all values. The one project underpins, and clears the way for, the other. In line with this distinction, Nietzsche identifies two connected tasks which need to be accomplished (*Genealogy*, Preface §6): "a *critique* of moral values" and a "knowledge of the conditions and circumstances under which they grew, evolved and changed" which is necessary for the critical task. The critique of morality is not something that is *effected* by means of genealogy: genealogy is, rather, an essential preparation for it (cf. Schacht 1994, 429). It is important that Nietzsche conceives genealogy in this light, since if he were attempting to infer a judgement of the present value of a thing from the alleged disreputable origins of that thing he would be involved in the genetic fallacy. In fact, he repeatedly points out this fallacy himself, so that it would be surprising if he flagrantly committed it throughout one of his major works.[5] He did, of course, expect to engender in his readers a suspicion towards the values whose genealogy he outlines, and, once the critique gets underway, the genealogy can be drawn on for support.[6] For instance, it will then become relevant that much of what we think of as a coherent system of moral valuation represents a tangle of contingently superimposed, inextricable confusions: the historical account lays bare some of these confusions. But the point remains that the genealogy is not yet the critique.

If genealogy, then, is neither argument nor method, but mere preparation for a critique of all values, what function *does* it have? In order to characterise the approach, it is useful to remember how Nietzsche differentiates himself

---

[3] References to genealogy as a "method" are rife in the Nietzsche literature. Despite the protestations of Geuss (1999, 1, 17), this commonplace has now become so entrenched as to be impossible to root out.

[4] For genealogy as having a "philosophical function," see Guay (2006).

[5] See Nietzsche, *The Gay Science*, §345: "Even if a morality has grown out of an error, the realization of this fact would not so much as touch the problem of its value." See also the following statements in Nietzsche's notebooks: "*Genesis* [*Entstehung*] and *critique* of moral valuations. The two *do not* coincide, as is easily believed" (Nietzsche, *Kritische Studienausgabe* 12, 2[131]); "The question about the origins [*Herkunft*] of our valuations and our tables of values [*Gütertäfeln*] in no way coincides with their critique, as is so often believed" (*KSA* 12, 2[189]).

[6] As Nietzsche goes on to say in a passage continuing from that just quoted (KSA 12, 2[189]): "[...] just as surely as the insight into some pudenda origo brings with it for our sensibility a diminution in the worth of the thing that has thus come into being, and prepares the way for a critical disposition and attitude towards the same."

from one of the targets of his polemic – those "genealogists" before him who had glimpsed the idea of telling a story about the acquisition of our values but had, according to him, badly botched its execution. Chief among such genealogists was Paul Rée, who had told a broadly Darwinian story about the origin of morality in his *The Origin of Our Moral Sensations* (1877). The approach there, one that Nietzsche himself had closely followed in his own *Human, All Too Human* (1878) and other texts, is now rebuked for lacking what Nietzsche calls "historical sense."[7] That is, it had proceeded by mere speculative hypotheses about the origination of morality which lacked psychological and historical plausibility and was injected with the moral prejudices of the present. What was needed was, rather, to investigate the "real history," as Nietzsche puts it, of our morality itself. Genealogy, as Nietzsche conceived it, was in fact nothing more nor less than "history, correctly practiced."[8] Nietzsche makes it clear in the Preface to the *Genealogy* that his aim is to replace mere hypotheses about the origins of morality with a "real history of morality"; he is interested in "the morality which has really existed, really been lived," or, again, "that which can be documented, which can really be ascertained, which has really existed, in short, the very long [...] hieroglyphic script of the human moral past" (Nietzsche 1998 [1887] §7). A Nietzschean genealogy does not represent an argument or a method, then, but a historical narrative.

Nietzsche, of course, would hardly have expended so much effort on giving a history of our morality if he thought it only served to invite the reader to cast doubt in a general manner on our moral values, in preparation for their critique. In that case, it would seem as if the historical detail was irrelevant: one might simply raise the question of the "value of our values," and proceed straight to the critique. However, the historical detail is often important. It is alleged by Nietzsche that punishment, for instance, does not have its origins in the administration of justice but in the infliction of cruelty. Originally, according to Nietzsche, punishment came about because the victim of a crime wanted to exact compensation from its perpetrator by the infliction of an equivalent cruelty on that perpetrator (see Nietzsche 1998). What is significant here is that accepting the truth of the genealogy will put into question the legitimacy of a preexisting justification, equally relying on a historical narrative, namely a story about the origination of punishment in some divine commands, or a contract formed between rational agents, etc. The clash between the genealogy and this preexisting story will be particularly relevant when it comes to

---

[7] See Nietzsche's enumeration of these earlier attempts in *On the Genealogy*, Preface §4. The approach in question comes in for criticism in Essay I, §§1–3.

[8] Nehamas (1985, 246 n. 1). This important insight has been endorsed by Geuss (1999, 17), and Leiter (2002, 180). Others, however, continue to regard genealogy as a fictional construction.

the critique stage. The value of punishment is not itself directly "debunked" or undermined by the telling of the story; however, the story, if it is true, can debunk the preexisting historical narrative. It will be left to the critique to sort out whether robbing punishment of this preexisting narrative will now leave it entirely without foundation (or whether an alternative justification for it can now be looked for elsewhere).

Whereas Nietzsche used his genealogies to prepare us for a critical mood in relation to our values, a more positive project has since also been carried out under the banner of "genealogy." A particularly ambitious example comes in the form of the "vindicatory genealogy" of the virtues of truthfulness offered by Bernard Williams in his *Truth and Truthfulness: An Essay in Genealogy* (2002). Williams tells us this is not "a book about Nietzsche, but it uses a method for which I have borrowed a name from him, *genealogy*, and I intend the association to be taken seriously" (Williams 2002, 13). Williams calls his genealogy "vindicatory" in conscious contrast to Nietzsche's genealogies, which he characterises as "subversive." For Williams, a "genealogical explanation of an outlook or set of values" is vindicatory if it strengthens our "confidence in them" (ibid., 283, n. 19). Williams's vindicatory genealogy is certainly not merely preparative. It wants to accomplish its own argumentative task.

Nietzsche began from a question about our present values, and was led by it to trace the way they had come into being in the past. As Raymond Geuss has put it, a Nietzschean genealogy is like a pedigree in reverse (see Geuss 1999, 1). A vindicatory genealogy starts from the opposite end, in Williams's case with a fictional "state of nature" capturing certain basic human needs and limitations and representing a pared-down form of language-using society. Williams then seeks to "derive" the virtues of truthfulness (which consist of "accuracy" and "sincerity") from the activities of information-sharing that take place in this basic, philosophically abstracted setup (see Williams 2002, 38). Ultimately, he seeks to establish that the need to operate by these virtues which is common to any social setup shows that any such setup must respect the notion of "truth," and thus that so should we.

Now, how can a *genealogy* accomplish Williams's task? If a genealogy is merely a narrative, it would seem surprising that it could ground the sort of claims Williams has in mind. Some have asked why a narrative is needed at all to do this. As Colin McGinn points out, for example, "[a] genealogy of this kind can be illuminating if it is unclear how a particular human faculty could have come about, but it is hard to see how it can work to vindicate a value. We can, after all, come to see how various vices might come about (for instance, stealing), but obviously this does nothing to justify them" (McGinn 2003). The question of whether truthtelling practices are *vindicated*, in other words, is an independent matter—the genealogy is not going to help us with this. However these practices may have come about, it is not this that will encourage us to endorse them, but something that has to come from outside the narrative.

Clearly Williams thinks that exhibiting how these virtues operate within the narrative *is* of significance to their vindication. Thus he shows that they play an essential role in the societies outlined by the narrative. But, even once we have seen the function of these virtues, Williams is open to the charge that he can only show the *instrumental* value these virtues have in these contexts, not their intrinsic value. Williams is obscure on this point. He thinks he can show that the value of truth must be intrinsic, not merely instrumental. But he takes it as a sufficient condition for something to have an intrinsic value that "first, it is necessary (or nearly necessary) for human purposes and needs that human beings should treat it as an intrinsic good; and, second, they can coherently treat it as an intrinsic good" (Williams 2002, 92). It is not clear what such "coherent treatment" consists in, but it would appear that if it is sufficient for something to be intrinsic that it is treated *as if it were* intrinsic, we are starting to lose our grip on the distinction between intrinsic and instrumental. The narrative that Williams tells, then, can only show the role that the virtues of truthfulness play in a society, but it cannot vindicate the manner in which they do so.

It appears that, in fact, Williams conceives of his work as having powerful argumentative force that goes well beyond what the narrative by itself can achieve. This is the force of transcendental argumentation.[9] This emerges if we consider whom Williams wants to convince that they cannot live without truth, those he calls the "deniers" of truth – people who think that the notion of truth is somehow dispensable. The strategy that is supposed to convince the "deniers" of Williams's thesis appears to be the following. Effectively, the overarching argument of *Truth and Truthfulness* is that in order to have the society we do, of the type in which we can have the discussion found in the book, we must already be taking seriously the "virtues of truth" (i.e., abiding by the constraints of "accuracy" and "sincerity"). And we can convince those whom Williams calls "deniers" of this since, even for them, the existence of this society at least is undeniable. To deny the societal framework would be to deny that the very discussion is taking place; and this would be, even from the deniers' point of view, an intolerable self-contradiction. Williams's overall argument can thus be characterised as seeking to establish the necessary conditions of something undeniable, just as Kant attempted to do when arguing that the existence of the external world is a necessary condition of the undeniable truth that we have experience of a certain kind.[10] Adopting the virtues of truth, in Williams's account, becomes a condition of the possibility of our society. This approach has the advantage of managing to engage

---

[9] Cf. Hartmann and Saar (2003), who characterise Williams's approach as "quasi-transcendental."
[10] See Kant's *Critique of Pure Reason*, Refutation of Idealism (B275–B279).

with the "deniers" of truth targeted by the book, just as a transcendental argument supposedly manages to engage the radical epistemic sceptic.[11]

I will leave it to one side whether Williams achieves this ambitious task. I only want to suggest that the additional philosophical work involved in Williams's project reveals his awareness that the narrative by itself cannot do the work of vindication. We have seen that a genealogy, as Nietzsche conceived it, can have a specifically delimited role in preparing the way for a critique of morality. Such a role is not available to vindicatory genealogy, since the vindication of an object cannot be effected merely by telling a narrative. A narrative is sufficient to perform the role Nietzsche envisaged for his genealogies, since here there is a potential for the narrative to dislodge or undermine some existing narrative which is thought to be appealed to by the values that will be subject to critique. An analogous situation does not exist for vindicatory genealogy: it would have to demonstrate in the first place that it had a special authority as a narrative. It would have to present something like a Hegelian chain of necessary development, if not a series of necessary conditions similar to that drawn on in a transcendental argument. Or it would have to impute to the first link in the chain some special properties that are such as to be passed on without loss to the succeeding members of the chain.

2 Evolution

I would now like to turn to arguments that draw on evolution in their premises and contain conclusions about morality. A distinction that at first sight resembles that between debunking genealogy and vindicatory genealogy has been drawn by Richard Joyce in relation to evolutionary arguments. Joyce distinguishes between the evolutionary vindication and the evolutionary debunking of morality (see Joyce 2006, chaps. 5, 6). In either case, an argument is made from facts about evolution to some claim about morality. In the former case, it is claimed that some conception of morality has been bolstered or legitimated; in the latter case, it is held to have been undermined. The overall strategy of such arguments, thus, is to derive a claim about morality from some facts held to be yielded by our understanding of evolution.

Several evolutionary models suggest themselves that can be drawn on here. These have been very well rehearsed, so they hardly need spelling out in de-

---

[11] The transcendental argument is here taken in the sense established by the work of P. F. Strawson and those following him. It is by no means clear that Kant was interested in refuting a sceptic of this type.

tail. However, we may mention three main candidates. First of all, the model of inclusive fitness demonstrates a way in which altruistic behaviour towards those genetically related to us can be selected for. A side-effect of this pattern of behaviour is that those in close physical proximity to us who are not related to us also benefit from our altruistic actions, since for the purposes of selection it is enough that physical proximity is a rough, not an exact guide, to relatedness. Secondly, "reciprocal altruism" provides a model of cooperative behaviour between unrelated individuals where altruistic acts are performed in the expectation of future reward. Thirdly, the emergence of "genuine altruism" may well have involved a mechanism of group selection: those tribes in which there was genuine altruistic behaviour did better than those who did not, and with whom they were in competition.

The details of these evolutionary explanations do not matter for our purposes. What they all have in common is that they tell a narrative about what our ancestors did in the past. Indeed, it has become a commonplace that the structure of a Darwinian explanation is essentially that of a historical narrative.[12] It is such historical narratives, then, that are being drawn on in evolutionary arguments about morality. Consider for example this vindicatory argument from Richmond Campbell:

(1) If having some morality (rather than none) enhances the life prospects of every group member, then having some morality is justified for each group member.

(2) The biological explanation for morality implies that having some morality (rather than none) enhances the life prospects of every group member.

Therefore,

(C) Having some morality (rather than none) is justified.
(Campbell 1996, 24)

The phrase "the biological explanation for morality" is a gloss of precisely such a historical narrative. This narrative says that cooperation between our ancestors caused the adaptive fitness of each of them to increase. There is, of course, a problem in equating enhancement of life prospects in this sense (adaptive fitness) with enhancing our life prospects now; Campbell, in trying to justify our morality, is presumably trying to show something more than that our morality helps us to make babies, just as it so helped our ancestors. As Campbell admits, "the biological explanation for the existence of morality is, of course, not directly about improving life prospects. The evolutionary story focuses instead on

---

[12] Mayr (1997, 64), Lewontin (1991, 143), Bock (1977, 853) and Nitecki and Nitecki (1992).

improving expected individual fitness" (ibid., 25). But leaving this to one side, we can see that a historical narrative is being relied on.

Something similar happens in the evolutionary vindication of morality proposed by Robert J. Richards.

(1) The evidence shows that evolution has constructed human beings to act for the community good.

(2) To act for the community good is what we mean by being moral.

(3) Human beings are moral beings.   (from 1, 2)

Thus:

(C) Each human being ought to act for the community good.
(Richards 1986, 289)

If we analyse premise (1), we find that a narrative account is again being relied on. Evolution is said to have "constructed" us to act for the community good. This claim may itself be false, since it is not clear that it was ever to our adaptive advantage to "act for the community good," only to demonstrate social behaviour which has later come to serve us well in acting for the community good. But be that as it may, the claim is that something has been of adaptive advantage to our ancestors over a period of time. Again, to present this claim is equivalent to presenting a historical narrative about our ancestors. While some such narrative is very likely to be true, it does nothing to tell us whether the phenomenon that has arisen through the narrative is justified. Such an inference would fall foul of the genetic fallacy, in precisely the manner Nietzsche sought to avoid.

The same considerations will apply to evolutionary arguments that seek to debunk morality; they, similarly, will rely on evolutionary explanations in their premises, and, again, these can be reduced to historical narratives. We might think that the narratives could have the same preparatory role for a critique that Nietzsche's genealogies have. Indeed, in some cases, that is, for some conceptions of morality, they may indeed perform this function. A useful way to illuminate this is to look at Darwin's own account of the emergence of the "moral sense" in *The Descent of Man* (1871) in relation to the contemporary reactions to that text.

Here, whether or not the account is experienced as having debunking potential, we will see, hinges on prior expectations about the nature of morality, just in the same way that the efficacy of Nietzsche's genealogy in preparing the way for his critical task relied on pre-existing conceptions in the minds of his audience. Darwin saw the development of the "moral sense" as going hand-in-hand with the development of man's mental powers, and as picking up where the development of the social instincts in our ape-like ancestors had left off. Darwin considered it "highly probable" that "any animal whatever, endowed with well-marked social instincts [...] would inevitably acquire a moral sense or conscience, as soon as its intellectual powers had become as well, or nearly as well developed, as in man" (Darwin 2004, 120-121). He outlines a four-stage development in which (1) a preexisting social instinct is consolidated by (2) the development of the mental faculties (allowing for memory and comparison of past experiences), (3) language and (4) habit.

Frances Cobbe responded to Darwin's account in her review of the *Descent* entitled "Darwinism in Morals." Cobbe noted the potentially revolutionary implications of Darwin's account for morality, claiming that, if Darwin is right, "our moral sense [has] come to us from a source commanding no special respect" (Cobbe 1872, 10). She draws special attention to two principles which Darwin embraces which were new in his exposition and had not been subscribed to in the previous discussion of Darwinism and morality by Herbert Spencer. One of these was

> his frank admission, that though another animal, if it became intelligent, would acquire *a* moral sense, yet that he sees no reason why its moral sense should be the same as ours, or lead it to attach the idea of right or wrong to the same actions. In extreme cases (such as that of bees), the moral sense, developed under the conditions of the hive, would, he thinks, impress it as a duty on sisters to murder their brothers. (Ibid.)

Cobbe's response suggests two ways in which Darwin's account could prepare the way for a debunking of a certain conception of morality. Firstly, she notes that on Darwin's account our morality has come to us from a source commanding no special authority. This common reaction in Darwin's own time reflects the widely held view that it was intrinsic to morality that it should issue from such a source—usually a divine source. There was something disturbing about the idea that the rudiments of our morality could have been handed to us by the apes if it was thought that we, as moral beings, were made in the image of God. It is precisely this expectation that Nietzsche plays on in a passage in *Daybreak*, where he refers to the "ape that is found to stand grinning at the portal of humanity" (Nietzsche 1997 [1881], §49). This image has the power to shock or unsettle purely because something other than an ape is expected to stand at the portal of humanity. For those who do not have such expectations, the power to shock is non-existent.

A second premonition of debunking force is suggested by Cobbe's point that our morality might have been radically different from what it in fact is. This point has been recently rehearsed by Ruse and Wilson.[13] The consequence that

---

[13] "It is easy to conceive of an alien intelligent species evolving rules its members consider highly moral but which are repugnant to human beings, such as cannibalism, incest, the love of darkness and decay, parricide, and the mutual eating of faeces." Ruse and Wilson (1986, 186).

can be drawn from this, and it is indeed the one that Ruse and Wilson draw, is that it invalidates a conception of morality that regards itself as universal in the sense of applying not just to *us* but also to some counterfactual other rational beings that we might have been if evolution had taken a different course. However, it is probably not correct to infer from the premise that our moral sense might have been different, and indeed radically different to the extent of a radical value inversion, that this moral sense should be somehow put in question, or that *our* distinction between right and wrong is now under threat. We might have to conclude that our conceptions of right and wrong are local to us as a species, rather than universal across, say, all potential rational beings. But that may well be simply a fact to accept and accustom ourselves to, with no harmful consequences. It may be that, indeed, the narrative of evolution can prepare the way for such a reassessment of the status of the moral claims we make. Certain philosophers would hardly be dismayed by this; others would be given pause for thought.[14]

To attempt to vindicate morality by appealing to evolutionary explanations, then, is to look in the wrong place. Attempting to draw such substantive conclusions from what are disguised historical narratives will open up just those pitfalls Nietzsche skilfully avoided. Again, an analysis of Nietzsche's strategy is useful in pointing out the limited power of a debunking strategy appealing to evolution. Only in the light of some specific preexisting expectations can the debunking strategy be effective.

References

Bock, W. (1977). Foundations and method of evolutionary classification. In *Major Patterns in Vertebrate Evolution*. M. Hecht *et al.* (eds.). New York: Plenum Press.
Campbell, R. (1996). Can Biology Make Ethics Objective? *Biology and Philosophy* 11: 21–31.
Cobbe, F. P. (1872). *Darwinism in Morals and Other Essays*. London: Williams and Norgate.
Darwin, C. (2004 [1879]). *The Descent of Man*. Harmondsworth: Penguin.
Emden, C. J. (2008). *Friedrich Nietzsche and the Politics of History*. Cambridge: Cambridge University Press.
Frankena, W. K. (1939). The Naturalistic Fallacy. *Mind* 48: 464–477.
Geuss, R. (1999). Nietzsche and Genealogy. *Morality, Culture, and History*. Cambridge: Cambridge University Press.
Guay, R. (2006). The Philosophical Function of Genealogy. In K. Ansell-Pearson (ed.). *A Companion to Nietzsche*. Oxford: Blackwell.

---

[14] A philosopher who would be little dismayed by this outcome, it seems to me, is Bernard Williams. See Williams (1985).

Hartmann, M. and Saar, M. (2003). Bernard Williams on Truth and Genealogy. *European Journal of Philosophy* 12: 386–398.

Leiter, B. (2002). *Nietzsche on Morality*. London: Routledge.

Lewontin, R. C. (1991). Facts and the Factitious in Natural Sciences. *Critical Inquiry* 18: 140–153.

Joyce, R. (2006). *The Evolution of Morality*. Cambridge, Mass.: The MIT Press.

McGinn, C. (2003). Isn't It the Truth? *New York Review of Books* 50 (6), April 10, 2003 (review of Williams 2002).

Mayr, E. (1997). *This is Biology: The Science of the Living World*. Cambridge, Mass.: Harvard University Press.

_____ (2010 [1983]). How to Carry Out the Adaptationist Program? In A. Rosenberg, and R. Arp (eds.). *Philosophy of Biology: an Anthology*. Chichester: Wiley-Blackwell.

Nehamas, A. (1985). *Nietzsche: Life as Literature*. Cambridge, Mass.: Harvard University Press.

Nietzsche, F. (1997 [1881]). *Daybreak*. Ed. M. Clark and B. Leiter, Trans. by R. J. Hollingdale. Cambridge: Cambridge University Press.

_____ (1998 [1887]). *On the Genealogy of Morality*. Trans. by M. Clark and A. J. Swensen. Indianapolis: Hackett.

Nitecki, M. H. and Nitecki, D. V. (eds.) (1992). *History and Evolution*. Albany, N.Y.: SUNY Press.

Richards, R. J. (1986). A Defense of Evolutionary Ethics. *Biology and Philosophy* 1: 265–293.

Richardson, J. (2004). *Nietzsche's New Darwinism*. Oxford: Oxford University Press.

Ruse, M. and Wilson, E. O. (1986). Moral Philosophy as Applied Science. *Philosophy* 61: 173–192.

Schacht, R. (1994). Of Morals and *Menschen*. In R. Schacht (ed.), *Nietzsche, Genealogy, Morality*. Berkeley, Cal.: University of California Press.

Williams, B. (1985). *Ethics and the Limits of Philosophy*. London: Fontana.

_____ (2002). *Truth and Truthfulness: An Essay in Genealogy*. Princeton: Princeton University Press.

17

Evolution and Moral Scepticism

Makoto Suzuki

1 Introduction: Two Versions of Evolutionary Scepticism

Several philosophers suggest arguments to the effect that because moral judgment (concept, capacity) is explained evolutionally, moral objectivity or moral realism is threatened (e.g. Ruse 1986, 254; Kitcher 2005, esp. 175-176; Joyce 2006, esp. Ch.6; and Street 2006, 109-166).[1] The basic flow of the argument goes like this:

(1) We have moral judgments (concepts, capacities) we do because of natural selection.

(2) However, no moral property or moral fact of the sort that realism assumes appears in the evolutionary explanation.

(C) Thus, some sort of moral scepticism is true.

What sort of moral scepticism is true? It depends on whom you talk to. Michael Ruse says that our moral judgments are not objectively true. Philip Kitcher says that they should be given a noncognitivist understanding. Richard Joyce says that moral judgments turn out to lack justification. Sharon Street's conclusion can be put conditionally: unless moral judgments are systematically mistaken, they either lack truth value or can be true or false only in anti-realistic terms.

This chapter defends moral objectivity and moral realism, especially its naturalistic variation, against such an evolutionary argument. I will focus on Richard Joyce's and Sharon Street's arguments, which respectively present the "requisite adaptationist" version and the "content adaptationist" version of the evolutionary argument in the most detailed and refined form. The "requisite adaptationist" is found on the claim that some cognitive trait (organ, capacity)

---

[1] Brosnan (2010) points out that Frances Power Cobbe (1871) has suggested this line of argument just after Darwin published his *The Descent of Man* and that Ruse and Wilson (1986, 186-187) started the contemporary discussion. Though E. O. Wilson occasionally hinted at his scepticism about moral objectivity in his earlier works (e.g., Wilson 1975), he did not clearly present an evolutionary argument there (see Kitcher 1985, 417-434 and Singer 1981, 68-72).

requisite to ethics is evolutionarily adapted irrespective of whether ethical facts have been present. The "content adaptationist" version is found on the claim that the contents of our ethical judgments, that is, what ethical judgments we make (and do not make), are heavily influenced by natural selection. The "requisite adaptationist" version and the "content adaptationist" version can be run independently of each other. The "requisite adaptationist" version can hold that the contents of ethical judgments are totally determined by, for example, environmental and cultural factors; and the "content adaptationist" version can hold that none of the traits, organs and capacities involved in ethical judgments is evolutionary adapted. By examining Joyce and Street's arguments, we can evaluate the prospect of the two versions of the evolutionary arguments.[2]

2 The "Requisite Adaptationist" Version of Evolutionary Scepticism: Richard Joyce's Argument

2.1 Exposition

Before I present and examine Joyce's argument, I would like to make two preliminary remarks. First, Joyce's argument assumes cognitivism, that is, moral judgments are really beliefs and not mere non-cognitive states.[3] Because I am trying to defend moral realism, which asserts that moral beliefs are beliefs, I grant this point.

Second, the argument is intended only to reveal the lack of justification for moral beliefs; it is not supposed to entail that moral beliefs are false. Joyce points out that knowledge of a belief's genealogy could show the belief to be false only if the belief implies a contrary genealogical story. Because moral beliefs are not generally about the origins of the beliefs, their genealogy does not directly show them to be false (Joyce 2006, 179-180).[4] Joyce himself

---

[2] Kevin Brosnan has given a general objection to evolutionary argument against moral realism (2010), which is partly influenced by Sober (1994, especially 112n4). Joyce criticizes Kitcher (2005) in Joyce (2006, 175-176). William A. Rottsschaefer and David Martinsen (1990) criticize Michael Ruse. Hallvard Lillehammer (2003) critically examines Ruse as well as biologists Richard Alexander and Robert Wright's evolutionary scepticism. I do not necessarily agree with these authors in the evaluation of the evolutionary scepticisms.

[3] Joyce argues against non-cognitivism in Joyce (2006, 53-57).

[4] In another place, Joyce argues for moral projectivism on the basis of the genealogy of moral beliefs (Joyce 2006, 4.4). Projectivism about a domain is often taken to imply that simple statements concerning the domain are all false. Thus, to be consistent, Joyce has to deny this implication of projectivism.

believes that simple moral beliefs are all false[5], but he does not hold that the evolutionary argument by itself gives this result.

Now let me introduce the basic line of Joyce's Argument (ibid., 181).

(1) If we discovered that a pill has disposed us to form beliefs involving a particular concept (that otherwise would not figure in our beliefs) regardless of whether the concept succeeds in denoting something in the world, it would render these beliefs unjustified (at least until positive evidence comes about).

(2) "Were it not for a certain social ancestry affecting our biology, the argument goes—we wouldn't have concepts like *obligation*, *virtue*, *property*, *desert*, and *fairness* at all" (emphasis original).

(3) The innate moral concepts have been generated (that is, selected-for) independently of whether or not they succeed in denoting properties in the world (ibid., 183).

Therefore:

(C) Due to the analogy between the above pill and natural selection, the above discovery about the evolutionary origin of morality renders our moral beliefs unjustified (at least until positive evidence comes about).

Note that premises 2 and 3 together constitute the "requisite adaptationist" claim that some traits central requisite for ethics, in this case moral concepts, are evolutionarily adapted irrespective of whether ethical facts have been present.

To illuminate Premise 1, Joyce tells the story of "the belief pill" about Napoleon:

> Suppose the imaginary belief pills [...] dispose you to form beliefs involving a particular concept—a concept that otherwise wouldn't figure in your beliefs [...] it's just a "Napoleon pill" that makes you form beliefs about Napoleon in general. Without this pill you would never have formed any beliefs about Napoleon at all [...] you discover beyond any doubt that you were slipped one of these pills a few years ago. Does this undermine all the beliefs you have concerning Napoleon? Of course it does. A belief is undermined if one of the concepts figuring in it is undermined (ibid., 181).

In this story, forming beliefs as the result of taking the pill is entirely independent of whether or not the facts necessary to render the beliefs true obtain in the world. This is why the beliefs can be undermined. Joyce argues that natural selection works like the belief pill. As Premise 3 says, our forming of moral beliefs as the result of natural selection is entirely independent of whether or not

---

[5] Joyce argues for error theory about morality in (2001).

moral facts obtain in the world. Even if no moral fact were present, it would still be systematically useful for our ancestors to form moral beliefs (ibid., Chapters 3-5, esp. 130-131). This is why moral beliefs are undermined (ibid., 182).

Note that, as Joyce admits, if he is correct, the evolutionary account of moral concepts crucially differs from that of other selected concepts (ibid., 182-183). For example, suppose, like moral concepts, we have the concept of child because of natural selection; otherwise we would not have beliefs involving the concept of child. In that case, however, unlike the case of moral concepts, the assumption that some child was present in the ancestral environment is necessary for an evolutionary account of the concept of child and related judgments. If there had been no child in the ancestral environment, it would not have been useful for our ancestors to form beliefs about children. This is why our beliefs about children are not undermined by the genealogy of the concept.

2.2 Doubts about Joyce's Argument

There are two doubts about Joyce's argument, one about Premise 2 and the other about Premise 3. First, are moral concepts the products of natural selection? Second, even given that, would moral concepts have been evolutionary adapted irrespective of whether they denoted any properties? I will examine these problems in order.

About Premise 2: Are Moral Concepts the Products of Natural Selection?

It is doubtful whether we have moral concepts because of natural selection. Because seemingly many capacities and dispositions that work in the moral domain are shared by primates (de Waal 1996) and hence have a long history, it can be somewhat plausible that they are triggered by natural selection. However, perhaps, humans have come to have moral concepts only recently in their history, and hence, not by natural selection.

Joyce hints at an argument that, if successful, would show that certain moral concepts, such as the concepts of being (morally) wrong and of desert, are innate, by which Joyce means being selected-for, being evolutionary adapted (Joyce 2006, 2). Joyce's argument goes like this.

(1) Certain emotions, such as guilt (and perhaps shame, indignation, and moral disgust), are innate (ibid., 97).

(2) These emotions necessarily include the representations expressed by moral concepts, for example, the thoughts that I [the agent herself] did something morally wrong and so deserve some punitive response (ibid., 67-68 and 91-104).

(C) Therefore, these moral concepts are innate.

The innateness of guilt is controversial. This is partly because, unlike anger, disgust, fear, joy, sadness, and surprise, it is not associated with a universal and distinguishable set of face expressions (See Ekman 2007, 217-222 for discussion). Further, as Joyce himself realizes, the second premise is problematic: "occasionally guilt involves non-moral transgressions" (Joyce 2006, 104). And, apparently, it is sometimes appropriate to have guilt over the transgression of epistemic norms, for example, not to believe merely on the basis of wishful thinking. If so, guilt does not essentially involve distinctively moral concepts.[6] The same might be true of shame and indignation. Perhaps, moral guilt, moral shame, moral indignation, and moral disgust necessarily involve certain moral concepts, but they are presumably not innate but are culturally constructed out of more basic emotions and moral thoughts.

Joyce also argues that even young children can make the moral/conventional distinction, and this is a reason to believe in the existence of the innate tendency to employ moral concepts (ibid., 136-137). However, whether young children can make the general moral/conventional distinction is empirically controversial (Kelly, Stich, Haley, Eng and Fessler 2007). Furthermore, psychological studies might rather suggest that people distinguish prohibitions against harm and other norms that are not underwritten by strong disgust (Prinz 2008b, 431). Because moral norms include more than prohibitions against harm and conventional norms include norms underwritten by strong disgust, people do not make the moral/conventional distinction in the normal sense. Further, we can perhaps explain our ability to make this distinction without recourse to an innate moral faculty. For example, difference in the ways parents teach two types of norms might make children respond to them in dissimilar ways (Prinz 2008b, 431-434; Sterelny 2008b, 20n6).

Joyce takes children's ability to make the moral/conventional distinction to be a piece of evidence that some development of moral cognition "exhibits an extremely reliable sequence, it gets underway remarkably early, its developmental pathway is distinct from the emergence of other skills, and its unfolding includes abrupt maturations." (Joyce 2006, 135) However, one can doubt whether the developmental pathway is distinct from the emergence of other skills, that is, whether moral development is not explained by the biological preparedness of more general cognitive traits like perceptual and emotional biases (Sterelny 2008a, 21-24; Sterelny 2008b, 15-19). Furthermore, even if this point were conceded, it would not follow that specifically moral concepts, rather than other moral cognitive traits like moral perception and imagination, are biologically prepared (let alone being evolutionarily adapted).

---

[6] Some people even claim that the emotions involve no evaluative concept (Gibbard 1990, Chapter 7).

If the use of moral concepts is ecumenical, then it is a (defeasible) piece of evidence for their innateness. Thus, Joyce also argues that the tendency to make judgments in moral terms is "ubiquitous and ancient," and "exists in virtually every human individual." (Joyce 2006, 135) I can concede that the tendency to make judgments in *evaluative* terms is shared throughout times and places, but the question is whether the tendency to use *distinctively moral* concepts is universal. The latter claim is questionable. Note that non-moral evaluative domains, such as aesthetics, prudence, epistemology, and law, can and often do share evaluative terms with morality. For example, law includes the words of "reason," "requirement"/"prescription," "prohibition," "permission," "obligation," "transgression"/"violation," "right"/"entitlement," "ownership," "desert"/"merit," "fairness," "justice," "compensation," "responsibili-ty," "blameworthiness," "justification," "excuse" and so on. Thus, even if someone's talk or writing contains these evaluative terms or their cognates, we cannot directly infer that he is using distinctively moral concepts. He might be using non-moral concepts or using yet undifferentiated evaluative concepts.

At a few points Joyce fails to appreciate this point.[7] First, Joyce regards precepts in the Egyptian Book of the Dead and in the Mesopotamian epic of Gilgamesh and the existence of ownership in ancient periods as the traces of morality (ibid., 134–135). However, this is too quick: the concepts of precepts and ownership in question might not be distinctively moral. Second, Joyce argues that people, even when very young, are better at checking the truth of deontic conditionals than at checking the truth of indicative conditionals, and this is a reason for thinking that the tendency to make moral judgments is innate (ibid., 135–136).[8] However, this is mistaken because children are better at checking even non-moral, prudential deontic conditionals (Prinz 2008a, 389). Third, Joyce argues that if people "don't already 'get' moral thinking, then it is a mystery what might be said" by others' moral assertions, such as "This is (morally) mine" (Joyce 2006, 138–139). However, the concepts of ownership, right, and so on appear in non-moral evaluative discourses. It seems that as far as people get these evaluative concepts, they can come to understand others' moral claims.[9]

---

[7] This is ironic because Joyce himself correctly points out that several attempts to vindicate morality on the basis of natural selection conflate some non-moral kind of normativity with moral one (Joyce 2006, 156–176).

[8] See Prinz (2008, 398) for the criticism of this interpretation of the experiment.

[9] Kim Sterelny points out another problem:
...there is a tension between accepting an adaptationist model of evolved moral cognition (as Joyce does) and arguing that moral concepts are strictly unlearnable by a trial and error process. For both are incremental, feedback-driven hill-climbing models of the construction of a capacity. One mechanisms operates within a generation, and the other over many generations. There is a structural similarity between learning and adaptive evolution:

Let me consider more carefully whether moral concepts are really brought about by natural selection. It seems that we have most of our concepts not because of natural selection: for example, presumably, we have the concepts of the things that are discovered or invented after prehistoric times not because of natural selection. Thus, suppose, as Bernard Williams argues, morality and its components are modern phenomena (Williams 1985, 174-196). Then, because ancient and medieval people do not engage in moral discourse, they do not make moral judgments. If so, presumably, moral concepts are not innate and not the products of natural selection.

As you might suspect, whether morality (as distinguished from other evaluative domains) and moral concepts are universal somewhat depends on how morality is defined. Partly because Williams characterizes morality in a very specific way, he takes it to be a modern institution.[10] Joyce himself lists several important characteristics of moral judgments, but it does not include two of the traditional descriptions of morality: universalizability and impartiality. R. M. Hare famously claims that if one's assertion qualifies as a moral assertion, she is committed to (re)phrase it or its ground in universal terms, avoiding both proper nouns and words with indexical or demonstrative aspects (Hare 1952). Joyce supposes the situation where a Yanomamö claims that it is permissible to kill "foreigners," where foreigner means anybody who has not descended from the blood of Periboriwa. Joyce apparently thinks that even if she refuses phrasing the claim or its ground in universal terms, we take it to be a moral claim (2006, 71-72). However, though I take her claim to be an evaluative claim, it is doubtful whether it counts as a moral one (Singer 2009). As for impartiality, Joyce does not even bother to explain why it fails to count as a characteristic of morality. However, it is intuitive that in some fundamental sense, morality takes into account all those potentially affected (Railton 2003, 360), or everyone's good is equally important morality-wise.[11] I wonder whether an evaluative system qualifies as moral if it does not satisfy such a condition: for it seems to

---

they are both trajectories of gradual improvement in response to success and failure signals from the environment. If the concept of a moral transgression is strictly unlearnable from experience by a mind without any moral concept, it is cannot evolve via a Darwinian trajectory from such a mind, either. (Sterelny 2008b, 20)

[10] See Chappell 2009, Section 2.

[11] Though so-called partialists question the impartiality of morality in narrower senses, it might be doubtful whether they intend to deny the condition in some more relaxed sense (See Jollimore 2008, Section 6). For example, Lawrence A. Blum, a representative partialist, says, "Finally, my argument is not meant to deny the fundamental moral truth in the notion that each person's good is as worthy of pursuits as is any other's [...] What I have argued is only that it is not properly reflected by the demand that the agent himself be equally concerned with the fostering of everyone's good." (1980, 66)

me that partly because morality is impartial, it is authoritative in the peculiar sense in which mere prudential or familial norms are not. However, if universalizability and impartiality are the characteristics of morality, it is doubtful whether morality existed until recently. Perhaps human beings had made evaluative judgments, but not moral judgments, until recently.

Though Joyce's list of the important characteristics of humans' moral judgments does not include impartiality, it still includes two conditions that concern the relationship between morality and prudence. "Moral Judgments pertaining to action purport to be deliberative considerations irrespective of the interests/ends of those whom they are directed; thus they are not pieces of prudential advice." (Joyce 2006, 70) "Moral judgments centrally govern interpersonal relations; they seem designed to combat rampant individualism in particular" (ibid.). If we take these statements to be the criteria of moral judgments, once again many people in the past might turn out to not make moral judgments. Ancient philosophers' ethical systems, for example, Aristotle's, emphasize acting both self-benefiting and other-benefiting virtues (Slote 1997, 195), including the cardinal virtues of justice, prudence, temperance, and fortitude. Medieval western societies continue this tendency illustrated by the list of seven deadly sins (pride, covetousness, lust, anger, gluttony, envy, and sloth) and by the list of cardinal virtues, which adds faith, hope, and charity to the ancient one. Joyce cites psychologists Tisak and Turiel (1984) and states that even young children are able to distinguish moral norms from prudential norms (Joyce 2006, 135). However, what Tisak and Turiel (1984) actually establish by their experiment is that young children evaluate interpersonal normative questions and intrapersonal normative questions differently. This result does not show that people are prepared by natural selection to make moral judgments in the narrow sense in which moral judgments centrally concern interpersonal problems. Thus, many people might not have made moral judgments in that sense, so according to the above standards, they might not be equipped with moral concepts. Presumably they are not innate or the products of selection.

Actually, Joyce is not committed to the view that the above characterizations are *conceptual* truths about moral judgments (Joyce 2006, 66 and 71). However, many contemporary people take morality to be conceptually distinct from prudential considerations and centrally concerned with interpersonal affairs. On their conception of morality, plausibly humans are not equipped with moral concepts by natural selection.

I suspect that acquiring moral concepts distinct from the concepts of tribal rules, law,[12] religion, and rationality might have occurred only recently. If so, it

---

[12] Joyce cites the psychological studies to show that cross-culturally, even young children distinguish—"moral" transgressions from merely "conventional' transgressions" (2006, 136). Even

is too short for natural selection pressure to select the distinctly moral concepts. About Premise 3: Is the Origin of Moral Concepts Different from That of Other Innate Concepts?

However, for the sake of argument, suppose that moral concepts are innate (i.e., evolutionary adapted). Now, Joyce's crucial assumption is that, unlike other such concepts, the reproductive advantage of moral concepts would have been the same whether or not they had denoted any properties. However, this assumption is questionable: moral concepts might have been useful precisely because they had denoted some properties. The hypothesis that moral concepts are selected-for because they denote certain real properties can be more plausible than Joyce's hypothesis that they are selected-for whether or not they denote any properties. Here, I will focus on the concepts of rightness and on virtue and vice concepts.

2.3 On the Concept of Rightness

Suppose, you are, like many naturalistic moral realists (Boyd 1988; Brink 1989; Railton 1986), sympathetic to utilitarians and think that if there is moral rightness, it is (at least to a large extent) determined by the well-being of those affected by actions. If each of us can and do (or each of our ancestors could and did) track what maximally promotes the well-being of those affected by actions and be prompted to act in that direction, then everyone tends to be helped: (given that the population is the same before and after the action) the average utility is maximized. Now, presumably, our ancestors' fitness positively correlates with their (and their important others') well-being. It is highly probable that they generally act to enhance the well-being. Thus, if the well-being is not positively correlated with their fitness, they will not likely to survive and have descendents. Because our ancestors survived and had descendents, probably the well-being is positively correlated with their fitness. Because the increase of the well-being has usually coincided with that of reproductive advantage, utilitarians can argue that having the concept of rightness has been reproductively advantageous at least partly because there is moral rightness and people can denote it by the concept. This is essentially the same as having the concept of child has been useful because there have been children and people can denote them by the concept.

---

if we suspend our qualms about this interpretation (see the text above), perhaps until modern ages, not only moral transgressions but also the transgressions of tribal rules are taken to be non-conventional. And until legal positivists like Jeremy Bentham (1782) and John Austin (1832) came around, laws might also have been taken by most to be non-conventional; think about the pre-modern natural law tradition led by, say, Thomas Aquinas.

Someone might object as follows. Acting on what is right would promote the total well-being but not always the agents' (and their kin's) well-being. Thus, even if the agents' fitness positively correlates with the agents' (and their kin's) well-being, having the denoting concept of rightness would not be reproductively advantageous for each agent. This objection assumes that the reproductive advantage comes (only) from directly acting on what is right. However, having the denoting concept of rightness would be reproductively advantageous in more indirect ways. For example, it is beneficial for each agent to have one another act on what promotes the total well-being, so it is beneficial to share the concept that denotes the property. By sharing the concept, they can focus on and discuss the topic of what maximally increases the well-being, and eventually create or change customs/conventions, educations/trainings, and the social system of punishment and reward so that each agent is prompted to promote the total well-being. In this way, having the denoting concept of rightness would be reproductively advantageous for an agent even if her acting always on what is right were not.

And both westerners and easterners (like me) have found somewhat attractive the idea that the morally right thing to do is promoting the well-being of everyone affected by the action. Thus, the hypothesis that the concept of rightness is selected-for because it denoted a certain real property (i.e., the maximization of well-being) can be more plausible than Joyce's hypothesis that it is selected-for whether or not it denoted any property. In fact, one can perhaps argue that our hardwired process of (though not exclusive to) moral judgment partly involves the tracking of such a real property. The neuropsychologist Joshua D. Greene amassed an impressive amount of psychological and neurophysiologic evidence that suggests we have the two processes of moral judgment, which sometimes tempt us in conflicting directions (Greene 2008). According to the view, while characteristically deontological judgments tend to be produced by emotional responses, characteristically consequentialist (utilitarian) judgments tend to be produced by more cognitive, emotionally neutral, processes.

> The only way to reach a distinctively consequentialist judgment ... is to actually go through the consequentialist, cost-benefit reasoning using one's "cognitive" faculties, the ones based in the dorsorateral prefrontal cortex. (Greene 2008, 65)

Greene thinks that this is because "[C]onsequentialism is, by its very nature, systematic and aggregative" (ibid., 64). So, one can perhaps argue that at least the utilitarian process of moral judgment is implemented by a hardwired tracking organ. We have some evidence for the hypothesis that the concept of rightness is selected-for because it denoted a real property, the maximization of well-being.

2.4 On Virtue and Vice Concepts

It is perhaps easier to argue that having some virtue and vice concepts has been useful (reproductively advantageous) at least partly because there are virtues and vices and people can denote them by the concepts.[13] This is because various virtues and vices are character traits. By having the concepts of virtues and vices, you might be able to track the character traits of people and predict how they tend to behave. If you can predict how they tend to behave, you can expect what benefit or harm comes from interacting with them. This information is surely useful. This is perhaps too quick, so now, I will illustrate the usefulness of virtue and vice concepts.

Suppose that we have the concept of kindness as a virtue. It seems that some people are kind and others are not. In this situation, we can denote kind people by the concept and track them. It would have been useful to be able to do so and track kind persons. If you can track kind persons, you can reap the benefits of their kind behaviour by interacting specifically with them. The same kinds of story can be told about the other concepts of other-regarding virtues, for example, the concepts of being compassionate, benevolent, generous, friendly, and trustworthiness.

Now suppose that we have the concept of being violent. It seems that some people are violent and others are not. And if so, we can denote people by the concept and track them. It would have been useful to be able to do so and track violent persons. If you can track violent persons, you can avoid the harm of their violent behaviour by avoiding interaction with them, unless it is necessary. The same kind of story can be told about the other concepts of other-regarding vices, for example, the concepts of being cruel, blood-thirsty, mean, too aggressive, malevolent, and insensitive.

Even in the cases of so-called self-regarding virtues and vices, it would be useful to have their concepts and be able to track them. In general, if you cooperatively interact with people with self-regarding virtues and get your interests to overlap with theirs, you might be helped by their behaviour, and if you competitively interact with these people and get your interests to conflict with theirs, you might suffer from their behaviour. For example, an industrious person is good to have as your business partner, but bad to have as your competitor. In contrast, if you cooperatively interact with people with self-regarding vices, you might be troubled by their behaviour, and if you competitively interact with these people, you might benefit from their behaviour. For example, a foolish person is bad to have as your business partner but good to have as your

---

[13] David Papineau points this out to me at the presentation of "Evolution and Moral Scepticism," Knowledge, Value, Evolution: An International Conference.

competitor. In addition, if you can track people with self-regarding virtues or vices, you can learn from their examples and benefit yourself. If you can find people with self-regarding virtues, you can benefit yourself by imitating their behaviour. If you can find people with self-regarding vices, you can avoid harm by avoiding what they do.[14]

Now it is intuitively plausible that virtues and vices are real and that it is useful to have their concepts. However, there are at least two possible sources of doubt: (1) situationism in psychology and (2) disagreement about the list of virtues and vices. Let me briefly comment on them.

According to situationism in psychology, there is no global character trait that is stable across various circumstances. Traditionally, virtues and vices are thought to be global character traits. For example, a kind person has a reliable tendency to act kindly, whatever mood she is in, whether her environment is noisy or smelly or whether she is in a hurry. If situationists are right, there is no virtue or vice as a global character trait. Controversy around situationism in psychology has not been settled completely (See Doris 2002 and Campbell, Meerschaert, and Chemero Manuscript). However, even if situationists are right and there is no global character trait, more situation-sensitive character traits can exist, and hence more situation-sensitive virtues and vices can be real. For example, even if there is no kindness simpliciter, a person can be kind unless being detracted (by her dark mood, icky environment, being in the middle of something, and so on).

Move on to disagreements on virtues and vices. It is often pointed out that people have different views about what count as virtues and vices (MacIntyre 1981, esp. Ch. 14). The potentially troublesome case for realists about virtues is that some people have had apparently conflicting views of what count as virtues. The most famous contrasting pair is that of the so-called pagan views and Christian views. According to pagan views, represented by Aristotle, Cicero, Machiavelli, Nietzsche, and Hume, virtues include beauty, strength, courage, magnanimity, and leadership. In contrast, Christian views, represented particularly by theologians, include humility, meekness, quietude, asceticism, and obedience. They seem to include incompatible sets of traits. Form this observation, it might be concluded that there is no real virtue.

However, of course, this is too quick. First of all, we can question whether these two sets of traits are really incompatible. Many virtues are expected to activate only in certain circumstances. For example, the virtues of leadership and magnanimity are expected to control the behaviour of a person when she interacts with a person who has a lower status. And the virtue of obedience is

---

[14] A more elaborate account of why virtues and the ability to detect them are selected-for is given, for example, in Miller (2008).

expected to control the behaviour of a person when she interacts with a person who has a higher status. Thus conceived, these two sets of virtues are not incompatible. In this way, many of the apparently incompatible lists of virtues are not really incompatible. Second, even if it turns out that some people really hold conflicting views of virtues, it does not imply that there is no fact of the matter about virtues. The view of the one side can be plain mistaken in the same way heliocentric theory turns out to be. Or it is just that the concepts of virtue and vices are as partially vague as many concepts are. Thus, apparently incompatible views of virtues do not show that virtues are not real.

2.5 Joyce's Retort

The above reply to Joyce suggests that we take moral rightness as the maximization of well-being of those concerned and virtues and vices as certain character traits within the natural world. Joyce argues that moral naturalism is doomed because it fails to capture the practical clout of morality, especially the point that moral prescriptions necessarily give reasons for acting to any person (2006, 6.3 and 6.4, esp. 193). As my reply is in the spirit of moral naturalism, it is probably the target of this retort. However, there are a few problems with this retort. First, not all moral judgments are prescriptions, and judgments about virtues and vices in particular are not prescriptions. For example, my judgments that some people are arrogant or that my wife is compassionate are not in themselves prescriptions. So, moral naturalists with respect to virtues and vices do not have to explain the peculiar authority of moral prescriptions. Second, whether moral prescriptions necessarily give reasons for acting to any person is highly controversial (e.g., Railton 1986; Brink 1989, 43ff). Moreover, this alleged authority of morality has long taken to be an independent source of moral scepticism (Mackie 1977 and Joyce 2001). Moral scepticism from evolutionary genealogy loses its importance if it is based on a controversial premise that, if true, can fuel scepticism on its own, without recourse to evolutionary history.

Thus, despite what Joyce argues, it is not shown that moral concepts are selected-for. And, even if this were the case, the hypothesis that moral concepts are selected-for because they denote certain real properties could be more plausible than Joyce's hypothesis that they are selected-for whether or not they denote any properties. Premises 2 and 3 are not shown to be true, so Joyce's "requisite adaptationist" version of the evolutionary argument fails to show that our moral beliefs are unjustified.

3 The "Content Adaptationist" Version of Evolutionary Scepticism: Sharon Street's Argument

3.1 Preliminary Remarks on Street's Argument

Next, I will consider Sharon Street's argument presented in her "A Darwinian Dilemma for Realist Theories of Value" (2006), which one author describes as "fascinating" (Lenman, 2008, Note 20).

Street's argument purports to be an objection against realist theories of value, which holds not only that there are evaluative truths but also that they are mind-independent, that is, *independent of some subject's (personal and subpersonal) evaluation or desire* (Street 2006, 110).[15] Street herself defends the view that evaluative truths are mind-dependent. I intend to defend not realism with regard to value in general but ethical realism, that is, realism specifically with regard to ethical values. So, I will critically examine Street's argument only in so far as it concerns ethical realism.

Keep in mind the target of Street's argument. She intends to refute realism in the sense that some evaluative judgments are true irrespective of some subject's evaluation or desire. On this understanding, the following views count as a version of anti-realism: straightforward non-cognitivism (all evaluative judgments lack truth values), error theory (no simple evaluative statement is true), dispositionalism or constructivism (the truth value of evaluative judgments depends on the actual or ideal evaluative attitudes or desires that some subject has). This narrow definition of realism is not universally adopted (even among metaethicists). Some might question whether the defense of realism in such a strong sense is necessary for the objectivity of value. However, increasingly, resurgent non-naturalist realists (e.g., Shafer-Landau 2003; Enoch 2007), Cornell naturalistic realists (e.g., Boyd 1988; Sturgeon 1985; Brink 1989), and Railton 1986 (about moral rightness), for instance, apparently hold realism in this strong sense. So, Street's argument is interesting as a potential objection against these metaethical positions.[16]

---

[15] More precisely, realism must also avoid achieving this independence by means of rigidifying the designating evaluative term/concept (Street 2006, 138–139). This complication does not affect the following argument.

[16] In another paper Street argues that her evolutionary argument also casts doubt on expressivist quasi-realism of the sort that Simon Blackburn (1993) and Alan Gibbard (2003) hold (Street forthcoming). This chapter does not examine this extension of the argument.

3.2 Exposition

The Starting Point of Street's Argument

Street holds that the ethical judgments we make are heavily influenced by the pressure of natural selection. Her evolutionary scepticism significantly differs from Joyce's "requisite adaptationist" version, which I have criticized above, particularly because her argument is founded on the "content adaptationist" view. Street holds this view partly because, across times, places, and cultures, the pattern of ethical judgments by humans is similarly fitness-enhancing.[17] For example, humans have the tendency to make the judgment (1)-(6), following which will be advantageous for survival and reproduction:

(1) The fact that something would promote one's survival is a reason *in favour of* it.

(2) The fact that something would promote the interests of a family member is a reason *to do* it.

(3) We have greater obligations to *help our own children than we do to help complete strangers*.

(4) The fact that someone has treated one well is a reason *to treat that person well in return*.

(5) The fact that that someone is altruistic is a reason *to admire, praise, and reward him or her*.

(6) The fact that someone has done one deliberate harm is a reason *to shun that person or seek his or her punishment*.

In contrast, humans have the tendency not to make the judgment (1')-(6'), following which will be disadvantageous for survival and reproduction:

(1') The fact that something would promote one's survival is a reason *against* it.

---

[17] The other alleged reason is that "the striking continuity that we observe between many of our own widely held evaluative judgments and the more basic evaluative tendencies of other animals, especially those most closely related to us." (Street 2006, 117) However, I think that empirical studies have not established that non-human animals have evaluative tendencies homologous to our tendencies to make evaluative judgments (cf. Prinz 2008, 397-402). And even if non-human animals have such evaluative tendencies, there will a methodological problem of how to identify their contents and compare them with the contents of our evaluative judgments so as to show the continuity or similarity.

(2') The fact that something would promote the interests of a family member is a reason *not to do* it.

(3') We have greater obligations to *help complete strangers than we do to help our own children*.

(4') The fact that someone has treated one well is a reason *to do that individual harm in return*.

(5') The fact that someone is altruistic is a reason *to dislike, condemn, and punish him or her*.

(6') The fact that someone has done one deliberate harm is a reason *to seek out that person's company and reward him or her*.

Thus, the ethical judgments we make are heavily influenced by the pressure of natural selection. This is the first premise of Street's argument, which I reconstruct below

The Basic Line of Sharon Street's Argument (2006, Section 5 and 6)

(1) The ethical judgments we make are heavily influenced by the pressure of natural selection.

(2) If there are mind-independent ethical truths, the pressure of natural selection either has or does not have some relation with the mind-independent truths.

(3) The pressure has no relation with the mind-independent truths, because any mind-independent truth could relate to the fitness of our ancestors only through their ability to track them, but it is implausible to hold that the capacity of ethical judgment was reproductively advantageous for ancestors because it is the ability to track ethical mind-independent truths.

(4) If the pressure has nothing to do with the mind-independent ethical truth, our ethical judgments are probably systematically mistaken.

(5) Such an estimated systematic error is implausible.

(6) There is no mind-independent ethical truth (because its assumed existence has an implausible implication).

Premise 3 needs explaining. Street points out that there are two alternative explanations of why our capacity to make certain evaluative judgments is selected-for. The first is the tracking account that because it enabled our ancestors to detect independent ethical truths, it enhanced their fitness. The second is the "adaptive link" account that because our capacity to make certain evaluative judgments "forged adaptive links between our ancestors' circumstances and their responses to their circumstances, getting them to act, feel, and believe in ways that turned out to be reproductively advantageous." (ibid., 127) Consider the judgment that the fact that something would promote one's survival is a reason to do it, the judgment that the fact that someone is kin is a reason to accord him or her special treatment, and the judgment that the fact that someone has harmed one is a reason to shun that person or retaliate. According to the tracking account, there are widespread human tendencies to make such judgments because making such evaluative judgments contributed to our ancestors' survival or reproduction because they are mind-independently true. According to the adaptive link account, in contrast, making such judgments contributed to reproductive success not because they are true but because they got our ancestors to respond to their environment with behaviour that itself promoted survival or reproduction. It tends to promote reproductive success to do what would promote one's survival, to accord one's kin special treatment, or to shun or retaliate those who would harm one (ibid., 128-129).

Street argues that because the tracking account is inferior to the adaptive link account, it is implausible to hold that we have ethical judgment because it enabled our ancestors to track ethical mind-independent truths. Any mind-independent truth could relate to the fitness of our ancestors only through their ability to track them. Thus, the pressure of natural selection has no relation with the mind-independent truth. This is what Premise 3 says.

Why is the tracking account inferior to the adaptive link account? According to Street, the tracking account is less parsimonious, is less clear, and answers fewer questions. The tracking account is less parsimonious because it refers to the mind-independent ethical truths that the adaptive link account does not (ibid., 129). The tracking account is less clear because it, without some addition, does not tell us exactly how it promoted our ancestors' reproductive success to grasp independent evaluative truths (ibid., 129-132). The tracking account fails to answer the three questions that the adoptive link account answers. First, it fails to explain the remarkable coincidence that so many truths it posits turn out to be the very same judgments we would expect to see if our judgments had been selected merely for them to be adaptive links between circumstances and response (ibid., 132). Second, the tracking account has trouble explaining why human beings have certain deep tendencies to make dubious (that is, presumably mistaken) judgments, such as judging in favour of those in the in-group over those in the out-group. If they are not true, the supposition that we have a tracking ability does not explain why we tend to make them (ibid., 133). Third, the adaptive link account informatively explains why we do not make certain ethical judgments (e.g., (1')-(6') above) by pointing out that making these judgments forges links between circumstances and responses that would have been reproductively useless or maladaptive. On the other hand, the tracking account just says that we do not make these judgments because they are false, which is not informative (ibid., 133).

As the conclusion of this argument, there is no mind-independent ethical truth. Street holds, however, that there are mind-dependent ethical truths. Then, why are mind-*dependent* ethical truths compatible with natural selection? Because the pressure of natural selection or the fitness of our ancestors can relate to mind-dependent ethical truths in a way other than their ability to track the truth. Suppose the mental processes of ethical judgment, say emotional capacities, have been fitness-enhancing because they independently help our ancestors survive or reproduce. Then, if ethical truths are largely determined by the outputs of the capacities, the truths are indirectly influenced by the pressure of natural selection. So, admitting such mind-dependent ethical truths does not entail that probably our ethical judgments are systematically mistaken (ibid., Section 10).

3.3 Doubts about Street's Argument

One can raise many doubts about Street's argument, especially Premises 3, 4, and 5. Let me examine these premises in the reverse order.

About Premise 5: Is the Prospect of Systematic Error in Ethical Judgments Really Implausible?

Premise 5 says that the prospect of systematic error in ethical judgments is implausible. However, this can be questioned. In many domains other than ethics (logic, statistics, physics, biology, psychology, and so on), the prospect of systematic error is seriously considered (Pinker 2002, Chapter 13). In fact, for example, it is well established that we systematically overestimate our own character and abilities (Hoffrage 2004). So, why can we be sure that the prospect of systematic error in ethical judgments is implausible?

In fact, there is a strong independent reason to think that people's ethical judgments are often mistaken. When our ethical judgment conflict with others' or our previous judgment, we naturally think that one of the conflicting judgments is mistaken. Given abundant ethical disagreement of this sort, we should perhaps expect that many of our current ethical judgments are mistaken.

Further, given a certain normative ethical theory, a systematic mistake has been predicted. For example, consequentialism (e.g., utilitarianism) often declares that our common intuitive judgments are false. In fact, most consequentialists have held that many commonsense moral judgments are systematically mistaken because they are agent-relative rather than agent-neutral. For example, commonsensically, each agent is required to help his or her child but not another's child (see (3) in Street's list). So, each agent has a different moral goal. Many consequentialists argue that this view is mistaken. Each person is required to help any child, whoever its parents are. This is because the revision, if followed by everyone, would better achieve the aim of each child being helped (e.g., Parfit 1984, Chapter 4). The point is that, depending on one's normative view, systematic mistake in commonsensical moral judgments is already expectable. For example, because a consequentialist is anyway ready to claim that the systematic mistake is present, she has nothing to lose by additionally holding moral realism and conceding the prospect of a systematic error.

Note that it is not only consequentialists who argue that there is a systematic error in commonsense ethical thinking. For example, Michael Slote, a representative virtue ethicist, claims "that our ordinary intuitive moral thought is not just complex, but subject to paradox and internal incoherence" (Slote 1997, 180). His defence of virtue ethics partly depends on the argument that it avoids such paradox and incoherence (ibid., 180-188).

Street probably responds to this line of objections as follows. It underestimates the depth of our systematic error she talks about. If the pressure of natural selection has nothing to do with mind-independent ethical truths, humans' moral judgments will be systematically mistaken not only now but forever: they will be irredeemable. Non-ethical judgments are just currently mistaken, and that might not be implausible. Some normative theories claim that many of our ethical judgments are currently mistaken, and this might not be implausible. However, it is implausible that our ethical judgments are systematically mistaken to the irredeemable extent. OK, then we will need to consider whether our ethical judgments will be in such a dire situation if the pressure of natural selection has nothing to do with mind-independent ethical truths. This will lead to us to the examination of Premise 4.

About Premise 4: If Natural Selection Has Nothing to Do with Mind-Independent Ethical Truth, Are Our Ethical Judgments Systematically Mistaken (Now and Forever)?

Premise 4 says that if the pressure of natural selection has nothing to do with mind-independent ethical truths, our ethical judgments are probably systematically mistaken. This is perhaps too quick; given that humans have the power of rational reflection. Even if we are led by natural selection to wrong conclusions, rational reflection might prevent us from going there. Street foresees and tries to rebut this response. Rational reflection about evaluative matters necessarily proceeds from some evaluative premise. If natural section contaminates our tendencies to make ethical judgments, then rational reflection becomes a process of assessing evaluative judgments that are mostly off the mark in terms of others that are mostly off the mark. And, this will not get us closer to evaluative truth (Street 2006, 124).

However, this argument might prove too much, and it might overlook the communicative and accumulative aspect of human reflection. In ancient times, our theories of the world were massively mistaken. Humans started to reflect on the basis of these incorrect theories of the world. This is partly due to the fact that, as cognitive scientists argue, humans are born with inaccurate folk theories of the world (Pinker 2002, Chapter 13). If Street were right, humans would probably have not improved their scientific knowledge. However, apparently, in the end they have come closer to the facts of the matter about what the elements, the structures, and the laws of the world are, overcoming the inaccurate folk theories.[18] Scientific theories have come closer to the truth partly because humans could and did talk about science and communicate the products of their reflection with one another. Because humans could and did talk about ethics and communicate the products of their reflection with one another, they might be able to come closer to mind-independent ethical truths even if natural selection disposes us to believe in an obstructing way.

This argument appeals to the analogy between ethics and science, so its plausibility depends on what sort of facts mind-independent ethical facts are. If ethical facts are non-natural facts, then this analogy is not strong particularly

---

[18] The above story presupposes that our current scientific theories or beliefs about unobservable entities and mechanisms are accurate. However, scientific anti-realists might object to this assumption, conceding that scientific theories have become more successful. They might argue, as Larry Laudan does (1981), that the history of science shows that, among successful scientific theories in the past, more of them turned out to be false to the extent that the theoretical entities they posited turned out to be non-existent. Thus, it is likely that more of the contemporary successful scientific theories will turn out to be false in the future to the extent that the theoretical entities they posit turn out to be non-existent.

However, scientific realists have several replies to this "pessimistic meta-induction." Firstly, scientific theories in the past are not as mature and successful as contemporary scientific theories. Thus, even if many of the former turn out to be false, many of the latter might be true. Secondly, the pessimistic meta-induction commits the turnover fallacy. Less accurate theories are more easily replaced than more accurate theories, so history tend to provide more instances of less accurate theories than more accurate theories. Moreover, even if more of the past successful theories turned out to be mistaken, we do not have to accept that more of the contemporary successful theories will follow the same course. That is because, thirdly, when the past theories are found to be mistaken, they are often replaced with the theories that correct the mistake. Thus, the replacing theories might well be more accurate. Furthermore, fourthly, even when past successful theories turned out to be mistaken, the replacing theories keep many aspects of the past theories. The success of the past theories then and the replacing theories now are best explained by the assumption that these maintained aspects are accurate. In these ways, scientific realists can defend the view that our current scientific theories or beliefs about unobservable entities and mechanisms are accurate.

because non-natural facts are presumably unreachable by observations, measurements, experiments, and other empirical or scientific methods. However, if ethical facts are natural facts, then this analogy is stronger. Again, the position of naturalistic realists is more secure.[19]

Street might object to the above argument by distinguishing the types of capacities involved in sciences and ethics. While the general cognitive mechanism engages in the production of scientific judgments, only a domain-specific mechanism engages in the production of ethical judgments (if the tracking account in ethics is true). She says at one point of her paper:

> The task of grasping independent evaluative truths presumably requires a highly specialized, sophisticated capacity, one specifically attuned to the evaluative truths in question. The capacity at issue is not a simple, brute sort of feature—not presumably, if we have any reasonable chance of grasping the truths posited by the realist. (Street 2006, 143)

A domain-specific mechanism is presumably a module, that is, the mechanism responding to a particular domain of stimuli, fast, automatic, and informationally encapsulated, that is, not much affected by information from other parts of the mind (Fodor 1983). Because of informational encapsulation, it is difficult to improve upon the outputs of the module. For example, the outputs of visual sense organ produce many illusions that the subject or the general cognitive mechanism knows to be false but cannot correct. For instance, in the Müller-Lyer Illusion, even after we know (say, by measurement with an accurate ruler) that the two lines have the same length, one line looks longer than the other. Thus, Street might argue, because only a domain-specific mechanism is involved in the production of our ethical judgments, that we cannot correct them.

However, first of all, it is unclear why ethical realists must take the faculty of evaluative judgments to be domain specific. It seems that unless evaluative

---

[19] Street suggests that naturalistic realists have trouble explaining how we can figure out the identity (or constitution) of normative properties with (or by) natural properties. She thinks that naturalistic realists' only data are our existing normative judgments. She argues that these judgments are heavily influenced by selection pressure, so our judgments about natural-normative identity (or constitution) pose the same problem for naturalistic realists as our first-order normative judgments do (Street 2006, 139–141). In the text I have been arguing that even if naturalistic realists' only data are our existing normative judgments, our first-order normative judgments will not pose the problem for naturalistic realists. Our judgments about natural-normative identity (or constitution) will not, either. And actually, naturalistic realists can use another data. Naturalistic realists can use empirical data to check whether the hypothesized natural-normative identity (or constitution) helps explain natural phenomena, for example, our moral learning, response, inference, judgment and behaviour (Railton 1986). Thus, naturalistic realists can argue that, partly by using the empirical data, we can come to figure out the natural-normative identity (or constitution).

facts are *sui generis* non-natural facts, realists do not have to posit a special mechanism for detecting these facts.[20] Second, even if there is a module for evaluation, it does not mean that its final verdicts cannot be challenged by a general cognitive mechanism. For example, in the Müller-Lyer Illusion, while the intermediate products, that is, the perception (the length of two lines looks different), are nearly impossible to correct, the final product, the judgment (the length of two lines are different), can be corrected. If the situation is analogous between vision and evaluation, though a general cognitive capacity might be unable to correct the intermediate productions, that is, ethical intuitions or gut reasons, it can change the final products of evaluation module, that is, evaluative judgments. And, Street has no argument to show that the general mechanism cannot be corrective in this way. To be sure, some psychologists, such as Jonathan Haidt, argue that pro-attitudes like emotions produce moral judgments, and our reasoning has little influence on our own moral judgments (Haidt 2001). However, Haidt's view is very controversial, and even Joshua Greene, who is very sympathetic to Haidt, holds that the general cognitive faculty can be involved in the production of ethical judgments (Greene 2008). And it appears the general cognitive faculty is involved in moral judgment. Both synchronically and diachronically, moral views vary culturally and individually to the extent that people have serious moral disagreement with one another. People sometimes engage in conscious reasoning and inference about moral issues, on the basis of which they sometimes change their ethical view.

Thus, it appears that naturalistic realists in ethics can deny Premise 4: if the pressure of natural selection has nothing to do with mind-independent ethical truth, our ethical judgments are probably systematically mistaken. Even if the antecedent is true, our ethical judgments might well fail to be systematically mistaken. And, even if many of our ethical judgments are now mistaken, they can be corrected in the future.

But do realists have to concede even the antecedent: the pressure of natural selection has nothing to do with mind-independent ethical truths? This question leads to the examination of Premise 3.

About Premise 3: Does Selection Pressure Have No Relation with the Mind-Independent Truth?

Premise 3 says that the pressure of natural selection has no relation with mind-independent truths, because any mind-independent truth could relate to the fitness of our ancestors only through their ability to track them, but it is implausible to hold that the capacity of ethical judgment was reproductively

---

[20] Prinz (2008a) and (2008b, 427–434) consider and reject various arguments for the existence of innate (specialized) moral mechanisms. See also Sterelny (2008a) and (2008b).

advantageous for ancestors because it is the ability to track ethical mind-independent truths. The argument for this premise is subject to several objections.

First of all, Street takes the tracking account and the adaptive link account to be mutually exclusive. However, this assumption can be questioned. Perhaps ethical judgments perform the double role of representing mind-independent truths and of thereby making the agent respond to the environment in a reproductively advantageous way. By analogy, think about physical pain. Physical pain might be a representation of a bodily part getting damaged, but it also makes the people who have it to respond in a reproductively advantageous way, forcing them to pay attention to the damaged bodily part.[21] If this sort of representational theory of physical pain is correct, apparently, both the tracking account and the adaptive link account are true of our capacity of having physical pain. In a similar way, both accounts might be true of our capacity of making certain ethical judgments. Because ethical judgments are often taken to have both cognitive and motivational functions, this is not a far-fetched possibility. Then, contrary to what Street assumes, the truth of the adaptive link account does not exclude the tracking account of our ethical judgment.

Suppose that, for the sake of argument, the tracking account and the adaptive link account are mutually exclusive. The second problem is that Street considers and compares the tracking account with the adaptive link account on a too abstract level. According to Street, the tracking account is less parsimonious, less clear, and answers fewer questions. However, the plausibility of this claim depends on what type of facts the specific tracking account takes ethical facts to be.

Street argues that the tracking account is less parsimonious because it refers to the mind-independent ethical truths that the adaptive link account does not. I admit that if the tracking account posits non-natural, *sui generis* ethical facts, then this is a big explanatory disadvantage for the tracking account. However, the tracking account is not committed to this view: it can hold that ethical facts are identical with or reducible to some naturalistic facts. In this case, no new kind of fact is introduced, so the tracking account is not so problematically extravagant. Perhaps the adaptive link account is still more parsimonious than the tracking account, but we should not prefer the former for that reason. To do so is, say, like preferring the non-representational theory of a mental state over the representational theory for the reason that the former does not refer to the represented fact. This is absurd.

---

[21] The presentational theory of pain is suggested, for example, by Dretske (1995, 102-103). If physical pain is the sort of the state I describe in the text, it is presumably a type of, what Millikan (2002) calls, "pushmi-pullyu representation": it not only descriptively represents one state of affair but also directively represents another state of affair. Millikan (2002) provides a naturalistic account of the pushmi-pullyu representation, especially in Chapter 6.

Street also argues that the tracking account is less clear because it, without some addition, does not tell us exactly how it promoted our ancestors' reproductive success to grasp independent evaluative truths. However, again this is too quick. Suppose, for example, that maximizing the well-being of those affected by action is what makes an action right. Then, the truth about rightness is independent of our evaluation or desire in the sense that the maximization is ethically correct whatever anyone thinks, feels, or desires. And, as I have argued concerning Joyce's argument, if this is what makes an action right, it might well have been beneficial for us and our ancestors to track the truths about rightness. As I also argued, virtues and vices have been reproductively advantageous to detect. Thus, depending on what types of facts the tracking account takes ethical facts to be, the account can be clearer by telling us how it promoted our ancestors' reproductive success to grasp independent evaluative truths.

Lastly, Street argues that the tracking account fails to answer the three questions that the adoptive link account answers. First, it fails to explain the remarkable coincidence that so many truths it posits turn out to be the very same judgments we would expect to see if our judgments had been selected merely for them to be adaptive links between circumstances and response. Second, the tracking account has trouble explaining why human beings have certain deep tendencies to make dubious judgments, such as judging in favour of those in the in-group over those in the out-group. If they are not true, the supposition that we have a tracking ability does not explain why we tend to make them. Third, the adaptive link account informatively explains why we do not make certain ethical judgments (e.g., (1')-(6') above) by pointing out that making these judgments forges links between circumstances and response that would have been reproductively useless or maladaptive. On the other hand, the tracking account just says that we do not make these judgments because they are false, which is not informative.

The first and the second points somewhat cancel out each other. Suppose, as the first point suggests, there is such a strong coincidence that many truths the tracking account posits are the same judgments we would expect to see if our judgments had been selected merely for them to be adaptive links. Then, humans will not have the innate tendencies to make dubious judgments because the innate tendencies lead to truths. If so, the tracking account does not have to explain why people have the tendencies to make dubious judgments. On the other hand, suppose, as the second point suggests, the adaptive link account needs to concede that some of our innate tendencies to make moral judgments are mistaken. Then, there will not be such a strong coincidence. So, the first point becomes less of an issue.

In fact, I think that the latter scenario is the case and that the coincidence is not so strong as to be problematic for the tracking account to explain. Many of the plausible ethical judgments are impartial. For example, everyone's wellbeing is (other things being equal) equally important; one should not harm any person; one should keep promise with any person; one should not deceive any person; every person has a dignity; helping any person is good; and it is wrong to use any person as mere means. It does not seem that these judgments or their application to cases would get or would have gotten the agent to respond in a reproductively successful way. I suspect that if there were reproductive advantages, they came only from the fact that these judgments were true (or close to truth). So, it is not the case that most ethical truths turn out to be the same judgments we would expect to see if our judgments had been selected merely for them to be adaptive links.

It is easier for the tracking account to explain why people tend to make certain dubious ethical judgments. As the adaptive link account says, it was reproductively useful to have these tendencies that make our ancestors to respond in certain ways. However, these judgments turn out to be mistaken. Because there are mind-independent ethical facts, representative error is always a possibility.

As for the third point, first of all, the supposed adaptive link explanation might go too far. It appears that if it makes it the case that we do not make ethical judgments (1')-(6'), it will also make it the case that we do not make impartial ethical judgments. This is because, seemingly, making impartial ethical judgments did not forge the adaptive links between circumstances and response that would be reproductively useful; impartial judgments would *not* be advantageous *for the responses or actions that the judgments enjoin*. However, we do make impartial ethical judgments. Because the adaptive link account might provide false (even if informative) explanations, it is too quick to conclude that the adaptive link account is explanatorily superior to the tracking account. Second, the tracking account can be combined with a plausible and informative explanation of why we do not make ethical judgments (1')-(6'). Probably due to natural selection, human beings are concerned more with the well-being of them and their kin than with that of strangers. Thus, human beings are not inclined to make ethical judgments (1')-(6'), which they see runs quite contrary to their concern. Thus, the tracking account together with a plausible psychological hypothesis can informatively account for why we do not make ethical judgments (1')-(6'). Once again, the tracking account can fare better than the adaptive link account.

Thus, naturalistic realists in ethics do not have to concede that natural selection has nothing to do with mind-independent ethical truths. Further, as I have argued earlier, even if this concession is made, they can deny the implication that our ethical judgments are systematically mistaken to an irredeemable extent. And, even if they are currently systematically mistaken, that is not implausible. Because Premises 3, 4 and 5 are not shown to be true, Street's "content adaptationist" version of the evolutionary argument fails to show that there is no mind-independent ethical truth.

4 Conclusion

Moral scepticisms from Joyce and Street's evolutionary arguments are not conclusive. Their arguments are significantly different: Joyce's argument represents the "requisite adaptationist" version, which is founded on the claim that some traits requisite to ethics, such as ethical concepts, are evolutionarily adapted irrespective of whether ethical facts have obtained. Street's argument represents the "content adaptationist" version, which is founded on the claim that the contents of ethical judgments are heavily influenced by natural selection. However, both arguments fail to justify a sceptical conclusion that our ethical beliefs are unjustified or that there is no mind-independent ethical truth. In particular, if you hold certain naturalistic realism, your position is not threatened by these evolutionary arguments: you are not caused more trouble from these arguments than (it is said) you already suffer elsewhere.

Acknowledgments

I first presented the previous Japanese version of this chapter titled "Is Normative Realism Compatible with Natural Selection?" at the 68th Conference of The Philosophical Association of Japan, May 16th, 2009, Keio University, Mita, Japan. I presented the revised English version titled "Evolution and Moral Scepticism" at "Knowledge, Value, Evolution: An International Conference on Cross-Pollination between Life Sciences and Philosophy," November 23–25th, 2009, Prague, Czech Republic. The current, final version of this chapter is substantially improved by the questions, comments, and criticisms given by the participants in these meetings. Needless to say, the remaining mistakes belong to the author and him alone.

References

Alexander, R. (1987). *The Biology of Moral Systems*. New York: Aldine de Gruyter.
Aquinas, T. *Summa Theologiae*.
Austin (1832). *The Province of Jurisprudence Determined*.
Bentham, J. (1782). *Of Laws in General*.
Blackburn, S. (1993). *Essays in Quasi-Realism*. New York: Oxford University Press.

Boyd, R. (1988). How to Be a Moral Realist. In Sayre-McCord, G. ed. *Essays on Moral Realism*. Ithaca: Cornell University Press, 181-228.

Blum, L. A. (1980). *Friendship, Altruism and Morality*. London: Routledge & Kegan Paul.

Brink, D. (1989). *Moral Realism and the Foundation of Ethics*. Cambridge: Cambridge University Press.

Brosman, K. (2010). Dissolving a Darwinian Dilemma for Moral Realism. A Presentation given in a History and Philosophy of Science Department Seminar at Cambridge University held on March 4, 2010.

Campbell, J. B., Meerschaert, S. and Chemero, A. (manuscript). *What Situationist Experiments Show*. http://edisk.fandm.edutony.chemerocampcoughchemsubmit.pdf

Chappel, T., Bernard Williams. In *The Stanford Encyclopedia of Philosophy*. E. Zalta (ed.). http://plato.stanford.edu/archives/win2009/entries/williams-bernard/

Cobbee, F. P. (1871). Darwinism in Morals. In *Darwinism in Morals: and Other Essays*. London: Williams and Norgate, 1872. http://www.archive.org/details/darwinisminmoral00cobbuoft.

Darwin, C. (1871). *The Descent of Man, and Selection in Relation to Sex*.

de Waal, F. (1996). *Good Natured: The Origins of Right and Wrong in Humans and Other Animals*. Cambridge, Mass.: Harvard University Press.

Doris, J. (2002). *Lack of Character: Personality and Moral Behavior*. Cambridge University Press.

Dretske, F. (1995). *Naturalizing the Mind*. Cambridge, Mass.: The MIT Press.

Ekman, P. (2007). *Emotions Revealed*. New York: Holt

Enoch, D. (2007). An Outline of an Argument for Robust Metanormative Realism. In *Oxford Studies in Metaethics* 2. R. Shafer-Landau (ed.). Oxford: Clarendon Press, 21-50.

Fodor, J. (1983). *The Modularity of Mind*. Cambridge, Mass.: The MIT Press.

Gibbard, A. (2003). *Thinking How to Live*. Cambridge, Mass.: Harvard University Press.

_____ (1990). *Wise Choices, Apt Feelings*. Cambridge, Mass.: Harvard University Press.

Greene, J. and Cohen, J. (2004). For the Law, Neuroscience Changes Nothing and Everything. *Philosophical Transactions of the Royal Society of London* (Series B, Biological Sciences) 359: 1775-1785.

_____ (2008). The Secret Joke of Kant's Soul. In *Moral Psychology*. Vol. 3: *The Neuroscience of Morality: Emotion, Brain Disorders, and Development*. W. Sinnott-Armstrong. Cambridge, Mass.: The MIT Press, 35-79.

Haidt, J. (2001). The Emotional Dog and Its Rational Tail: A Social Intuitionist Approach to Moral Judgment. *Psychological Review* 108 (4): 814-834.

Hare, R. M. (1952). *The Language of Morals*. Oxford: Oxford University Press.

Hoffrage, U. (2004). Overconfidence. In *Cognitive Illusions: a Handbook on Fallacies and Biases in Thinking, Judgment and Memory*. R. Pohl (ed.). New York: Psychology Press.

Jollimore, T. (2008). Impartiality. In *The Stanford Encyclopedia of Philosophy*. E. Zalta (ed.). http://plato.stanford.edu/archives/fall2008/entries/impartiality/.
Joyce, R. (2001). *The Myth of Morality*. Cambridge: Cambridge University Press.
_____ (2006). *The Evolution of Morality*. Cambridge, Mass.: The MIT Press.
Kelly, D., S. Stich, K. J. Haley, Eng, S. J. and Fessler, D. M. (2007). Harm, Affect, and the Moral/Conventional Distinction. *Mind & Language* 22 (2): 117-131.
Kitcher, P. (1985). *Vaulting Ambition*. Cambridge, Mass.: The MIT Press.
_____ (2005). Biology and Ethics. In *The Oxford Handbook of Ethical* Theory. D. Copp (ed.). Oxford: Oxford University Press, 163-185.
Laudan, L. (1981). A Confutation of Convergent Realism. *Philosophy of Science* 48: 19-49.
Lenman, J. (2008). Moral Naturalism. In *The Stanford Encyclopedia of Philosophy*. E. Zalta (ed.). http://plato.stanford.edu/archives/win2008/entries/naturalism-moral/.
Lillehammer, H. (2003). Debunking Morality: Evolutionary Naturalism and Moral Error Theory. *Biology and Philosophy* 18: 561-581.
Mackie, J. L. (1977). *Ethics: Inventing Right and Wrong*. London: Penguin Book.
MacIntyre, A. (1981). *After Virtue*. London: Duckworth.
Miller, G. (2008). Kindness, Fidelity, and Other Sexually Selected Virtues. In Sinnott-Armstrong (2008), 209-243.
Millikan, R. G. (2004). *Varieties of Meaning*. Cambridge, Mass.: The MIT Press.
Parfit, D. (1984). *Reasons and Persons*. Oxford: Clarendon Press.
Pinker, S. (2002). *The Blank Slate*. London: Penguin.
Prinz, J.J. (2008a). Is Morality Innate? In Sinnott-Armstrong (2008), 367-406.
_____ (2008b.) Reply to Dwyer and Tiberius. In Sinnott-Amstrong (2008), 427-439.
Railton, P. (2003). *Facts, Values, and Norms*. Cambridge: Cambridge University Press.
_____ (1986). Moral Realism. *Philosophical Review* 95 (2): 163-207.
Rottsschaefer, W.A. and Martinsen, D. (1990). Really Taking Darwin Seriously. *Biology and Philosophy* 5: 149-173.
Ruse, M. (1986). *Taking Darwin Seriously*. Amherst, N.Y.: Prometheus Books.
Ruse, M. and Wilson, E. (1986). Moral Philosophy as Applied Science. *Philosophy* 61: 173-192.
Singer, P. (1981). *The Expanding Circle: Ethics and Sociobiology*. Oxford: Clarendon Press.
_____ (2009). Richard Joyce, *The Evolution of Morality*. *Notre Dame Philosophical Reviews*. http://ndpr.nd.edu/review.cfm?id=6383.
Sinnott-Armstrong, W. (ed.) (2008). *Moral Psychology Volume 1: The Evolution of Morality: Adaptations and Innateness*. Cambridge, Mass.: The MIT Press.
Slote, M. (1997). Virtue Ethics. In M. W. Baron, P. Pettit and M. Slote. *Three Methods of Ethics*. Oxford: Blackwell Publishers, 175-238.
Sober, E. (1994). Prospects for an Evolutionary Ethics. In *From A Biological Point of View*. Cambridge: Cambridge University Press, 93-113.
Sterelny, K. (2008a). Moral Nativism: A Sceptical Response. *The Jean Nicod Lectures 2008: The Fate of the Third Chimpanzee*, Section 2. http://www.institutnicod.org/Session_2.pdf.

___ (2008b). Moral Nativism: A Sceptical Response: Background to Lecture 2. *The Jean Nicod Lectures 2008: The Fate of the Third Chimpanzee.* http://www.institutnicod.org/Backgroundto_Lecture2.pdf.

Street, S. (forthcoming). Mind-Independence without the Mystery: Why Quasi-Realists Can't Have It Both Ways. Nous.

___ (2008). Constructivism about Reasons. In *Oxford Studies in Metaethics* 3. R. Shafer-Landau (ed.). Oxford: Clarendon Press, 207-246.

___ (2006). A Darwinian Dilemma for Realist Theories of Value. *Philosophical Studies* 127: 109-166.

Tisak, M. S. and Turiel, E. (1984). Children's Conceptions of Moral and Prudential Rules. *Child Development* 55: 1030-1039.

Williams, B. (1985). *Ethics and the Limits of Philosophy.* Cambridge, Mass.: Harvard University Press.

Wilson, E. O. (1975). *Sociobiology: The New Synthesis.* Cambridge, Mass.: Harvard University Press.

Wright, R. (1994). The *Moral Animal.* London: Little, Brown & Company.

18

Evolutionary Origins of the Sense of Justice[1]

Wojciech Załuski

1 Three Perspectives for Analysing the Origins of Our Moral Tendencies

The question of the origins of our moral tendencies—one of the crucial questions of moral psychology—can be tackled from three different perspectives: purely biological, purely sociological, and a mixed one—biological-sociological. Each of these perspectives relies on specific assumptions.

The purely biological perspective assumes that our moral tendencies are innate—deeply embedded in our nature as a result of biological processes—and that thereby nothing really important is added to them in the process of socialization. Accordingly, this perspective assumes that human nature is essentially good. By contrast, the purely sociological perspective assumes that our moral tendencies arise only in the process of socialization. The assumption underlying this perspective may be either that human nature is deeply flawed, so that human innate antisocial tendencies have to be counteracted in the process of socialization, or that human beings are born with no morally relevant tendencies whatsoever—neither moral nor immoral—so that their moral tendencies have to be developed in the process of socialization. As we can see, the purely sociological perspective has two varieties: the pessimistic one, which assumes that the human mind is not a blank slate but is equipped with antisocial tendencies, and the more optimistic one, which assumes that the human mind is a blank slate upon which everything has to be written and can be written with equal ease. There is little to be said in favour of the purely biological perspective. By denying the role of the process of socialization in the development of our moral tendencies, this perspective is too evidently at odds with our commonsense knowledge about human nature. However, the other extreme perspective—the purely sociological one—is also, though less obviously, implausible. Even though many notable thinkers professed one of the views about human nature underlying this perspective, i.e., the view that human nature is deeply flawed or

---

[1] Some parts of this article are borrowed in extenso from the chapter IV of Załuski (2009).

the view that the human mind is a blank slate (or *quasi*-blank slate), these views seem to be untenable in the light of results of various biological sciences. One can give many biological arguments for the claim that human beings are born with a number of moral predispositions. These arguments come, e.g., from the evolutionary theory, which says that empathy, kin altruism, reciprocal altruism are biological adaptations, from primatology, which ascertains the existence of various moral predispositions in our closest relatives—nonhuman primates, and from neurobiology, which teaches us that moral-decision making involves evolutionary old—"emotional"—parts of our brains.

Thus, rather unsurprisingly, what seems to be the correct perspective for the analysis of our moral tendencies is the mixed one. It is therefore within this framework that I shall provide an analysis of one of such tendencies, namely, our sense of justice. Thus, the analysis is based on and develops the insight that our sense of justice can be decomposed into two different types: predispositions for the sense of justice, which can be explained on purely biological grounds, and a full-blown sense of justice, which cannot be explained on purely biological grounds. I shall call the former type "the rudimentary sense of justice" and the latter type "the genuine sense of justice."

Thus, I shall argue that only the rudimentary sense of justice is a biological adaptation, i.e., it was preserved by natural selection, as it increased the chances of survival and reproductive success of those who were endowed with it. I shall also argue that the rudimentary sense of justice is "Janus-faced," i.e., it is rational-emotional in character—it is greed constrained by our capacity to anticipate the reactions of other people *plus* a bundle of emotions: envy, the instinct for retaliation, gratitude, forgiveness, and the sense of guilt.[2] Before I present a more detailed account of these two types of justice, I shall present some elementary distinctions regarding the very concept of justice, because the types of justice are defined with reference to the concept of justice.

2 The Concept of Justice

One can distinguish two basic varieties of the concept of justice: material justice and retributive justice. Material justice can refer to two different situational contexts: the context of distributing goods or burdens among many persons (in this I call material justice "distributive justice") and the context of exchanging goods and burdens between two persons (in this context material I call

---

[2] It is to be stressed that the idea that these emotions play an important part in developing our sense of justice was first put forward by Trivers (1971).

material justice "exchange-regulating justice"³). The principles of distributive justice may take various forms, e.g., "to each according to her needs," "to each according to her desert," "to each an equal share." The principles of exchange-regulating justice are, e.g., "to each according to what she has agreed upon" or "the goods being exchanged should be of equal value." The second variety of the concept of justice is retributive justice. The principles of retributive justice specify what punishment should be imposed on persons who violated the requirements of just distribution prescribed by distributive justice and the requirements of just exchange prescribed by exchange-regulating justice (the clause *"inter alia"*is justified because retributive justice refers not only to the infringements of principles of justice, *e.g.*, to cheating in exchanges, but also to the infringements of other moral norms, *e.g.*, the norm which prohibits doing physical harm to other people or the norm which prohibits taking their property). The basic principle of retributive justice is "to each such a punishment which corresponds to the level of her guilt and the seriousness of harm which she has caused by her action." The common name "justice" for the otherwise different principles of material justice and retributive justice is justified by the fact that all these principles can be regarded as the specifications of the classical definition of justice proposed by the Roman lawyer Ulpianus (based on Aristotle's analysis in *Nicomachean Ethics*), according to which justice consists in granting each person her due (*suum cuique tribuere*).⁴

3 Two Pure Types of the Sense of Justice

By the genuine sense of justice I understand a virtue, i.e., a trait of human character, which embraces a cognitive element—the clear awareness of sophisticated principles of justice *qua* principles of justice, and a motivational element—the disposition to comply with these principles for the sole reason that they are principles of justice, i.e., out of pure respect for justice. By the rudimentary sense of justice I understand a virtue, i.e. a trait of human character, which embraces a cognitive element—the dim awareness of basic principles of justice (not necessarily *qua* principles of justice), and motivational element— the disposition to comply with these principles for other reasons than pure respect for justice. The principles of justice accepted by the agent endowed with the rudimentary sense of justice are simple, not sophisticated, not clearly

---

³ I avoid the traditional name "commutative justice," as it (on some accounts) embraces also what I call below "retributive justice."

⁴ The definition says precisely: *Iustitia est constans et perpetua voluntas ius suum cuique tribuendi* (justice is a constant and perpetual will to grant to each person her own right).

articulated, while the principles of justice accepted by the agent endowed with the genuine sense of justice are complex, sophisticated, clearly articulated. The agent endowed with the rudimentary sense of justice is motivated to act justly by egoistic motives and/or by certain emotions, while the agent endowed with the genuine sense of justice is motivated to act justly by the very content of the principles of justice. The rudimentary sense of justice is not a phenomenon *sui generis*, as it can be decomposed into its simpler, constituent elements, while the genuine sense of justice is a phenomenon *sui generis*. Of course, even though there are important differences between both forms of the sense of justice it may be impossible in many circumstances to state if a just action is a manifestation of the rudimentary sense of justice or of the genuine sense of justice. For instance, the fact that an agent who has received unjustly more goods than others decides to share her goods with the others may just as well be a manifestation of her genuine sense of justice as of her rudimentary sense of justice (e.g., she may fear the negative consequences of the others' envy). Clearly, both forms of the sense of justice are just pure types: one can distinguish some intermediate forms between them which combine somehow the elements of the genuine sense of justice and the rudimentary sense of justice.

4 The Rudimentary Sense of Justice as a Biological Adaptation

As mentioned, I shall argue for two main theses: first, that the rudimentary sense of justice is a biological adaptation, i.e., that it was preserved by natural selection, as it increased the chances of survival and reproductive success of those who were endowed with it; second, that evolutionary theory suggests that the rudimentary sense of justice is "Janus-faced," i.e., it is rational-emotional in character—it is constrained greed coupled with a bundle of emotions: envy, the instinct for retaliation, gratitude, and the sense of guilt. The second thesis does not imply, however, that constrained greed and the above mentioned emotions always act simultaneously. Rather, they constitute a group of "mechanisms" that underlie various forms of the rudimentary sense of justice manifesting themselves in various contexts (this is exactly what is implied by saying that the rudimentary sense of justice as understood here is "a pure type"). As can be readily seen, my account of the rudimentary sense of justice will be naturalistic not only in the sense that it assumes that this form of the sense of justice can be explained by appealing only to scientific methods (and thereby without assuming that it has been implanted in us by God or without positing that it constitutes some mysterious faculty to perceive moral facts) but also in the sense that it assumes that this form constitutes an evolutionary adaptation (and not some accidental by-product of evolution). By contrast, my account of the genuine sense of justice is naturalistic only in the former sense, as it assumes that even though the "transition" from the stage of the rudimentary sense of justice to the stage of the genuine sense of justice can be explained by scientific methods, the genuine sense of justice is not a biological adaptation but, rather, a manifestation of our capacity for abstract thinking.

I shall now turn to presenting the evolutionary account of the rudimentary sense of justice. My claim is that it embraces two elements: *constrained greed and a bundle of emotions*. I shall argue that these "mechanisms" tend to lead to *just actions*, which is why, their heterogeneity notwithstanding, they can be construed as constituting a (rudimentary) form of the sense of justice.

4.1 Constrained Greed

The first element of the rudimentary sense of justice is greed constrained by the capacity to anticipate reactions of other people. It seems plausible to argue that natural selection would favour people displaying constrained greed, i.e., a tendency to maximize their goods moderated by the cognisance of the fact that exceedingly aggressive pursuing of goods could engender a negative reaction of others and in consequence preclude achieving these goods. It would therefore favour those people who, while pursuing their own interests, were able to take into account the interests of others. Constrained greed generates various behaviours in various contexts of justice. In the context of distributive justice and exchange-regulating justice it functions in the following way. An agent who is greedy, i.e., desires to get the whole of a given good for herself, but simultaneously reasonable, i.e., cognizant of the fact that the realization of her desire is not possible (as this desire is very likely to engender a negative reaction on the part of the other potential beneficiaries of the good and, consequently, lead to her receiving no part of the good at all) will arguably act in accordance with the norm prescribing equal division. This is the most "obvious" or "salient" principle of justice, as it enables each person to receive as large a part of the good as it is possible and compatible with the same parts for the other persons. Arguably, apart from egalitarian principles of justice, constrained greed may also generate simple meritorious principles of justice directed against potential free-riders. In the context of retributive justice constrained greed functions in a different way. An agent who is greedy, i.e., wants to avoid the costs of punishing the wrongdoer, but is simultaneously cognizant of the fact that not punishing the wrongdoer may encourage her to do subsequent acts of wrongdoing with regard to the agent and thereby make the agent sustain larger costs than the costs of punishing the wrongdoer, is likely to act in accordance with some principle of retributive justice. As it seems, then, greed constrained by the capacity to anticipate the reactions of other people to unconstrained greed and to compare the potential consequences of this reaction with the consequences of constraining greed is sufficient to generate many kinds of just actions.

4.2 A Bundle of Emotions

The account of the rudimentary sense of justice presented in the previous is not complete, as evolution seems to have worked in a more complicated or, rather, more "cautious" way: it supplemented the fragile rational mechanism of constrained greed with a bundle of emotions which serve the same evolutionary goals as this mechanism. In other words, an agent's mere rational calculations, if carried out in a correct way, would suffice for her to reap maximum benefits from cooperative interactions. The problem is that humans cannot be expected to consistently carry out such calculations in a correct way. This explains in a general way why natural selection buttressed constrained greed by various natural emotions: envy, the instinct for retaliation, gratitude, the sense of guilt. These emotions supplement, correct, or substitute for such calculations, thereby moving an agent to undertake actions which she would have undertaken were she able to carry out all the necessary calculations for reaping possibly high benefits from cooperative interactions. These emotions, then, support constrained greed in realizing evolutionary aims, i.e., direct human behaviour in ways that were adaptive over evolutionary time. I do not claim that the above list of emotions is complete (arguably, it could be supplemented by other emotions, e.g., by forgiveness), that a person must display them all to be said to possess the rudimentary sense of justice, or that they all have to operate simultaneously. I claim that they are just main emotions with which evolution equipped us in order to reap maximum gains from cooperative interactions and which constitute typical elements of our rudimentary sense of justice. I shall now examine these emotions in greater detail.

Envy

Unlike greed, which is a two-place relation (it embraces a subject–a greedy person–and a given good), envy is a three-place relation–it embraces a subject (an envier), a rival (a party who is envied) and a good (*e.g.*, some possession, capacity, trait that the subject supposes the rival to have, or a particular person's affections directed toward the rival) (see D'Arms 2008). Envy is a manifestation of the human tendency to evaluate one's own situation in a comparative way, i.e., by referring it to the situation of other people. Accordingly, it is an evidence of the fact that people are concerned not with their absolute level of goods but with the relative level (i.e., compared with the standing of others). I shall present now three basic forms of envy—benign (admiring) envy, invidious (malicious) envy and temperate envy[5]—and reflect on which of them is part of the rudimentary sense of justice.

Benign envy is simply an unpleasant emotion (a distress, pain, nuisance, etc.) felt by the subject at the thought that she does not possess the good and the rival does, and unaccompanied by any kind of desire that the rival lose this good. As D'Arms points out, benign envy is difficult to distinguish from a positive desire for a good, i.e., from greed, or from admiration for the rival (cf. D'Arms 2008). Accordingly, benign envy cannot be the form of envy that I claim to be part of the rudimentary sense of justice in addition to constrained greed.

Invidious envy contains two elements: (1) an unpleasant emotion (a distress, pain, nuisance, etc.) felt by the subject at the thought that she does not possess the good and the rival does; (2) a desire that the rival lose the good—the desire *which appears despite the fact that it was not possible to distribute the good in such a way that the subject would receive a reasonable part of it.*[6] It seems that invidious envy is part of the psychological equipment of the malicious person. It is very dubious to contend that invidious envy forms part of the rudimentary sense of justice. Invidious envy—one of the greatest pathologies of the human spirit—is not only ignoble and vicious to feel (which, of course, would not by itself imply that its existence is not probable in the light of evolutionary theory) but also does not seem to bring any evolutionary advantage, as it is a highly self-destructive emotion. It seems to me that the type of envy that is part of the rudimentary sense of justice is what I call "temperate envy."

Temperate envy contains two elements: (1) an unpleasant emotion (a distress, pain, nuisance etc.) felt by the subject at the thought that she does not possess the good and the rival does; (2) a desire that the rival lose good—the desire *which appears because of the fact that it was possible to distribute the good in such a way that the subject would receive a reasonable part of it.* It is worth noting that while benign envy is less reprehensible than temperate envy (as it does not include the desire that the rival not have the good), temperate envy is less reprehensible than invidious envy (as it includes the desire that the rival not have the good only if it were possible to make an equitable distribution of this good). Unlike invidious envy, temperate envy seems to bring evolutionary advantages. It is a manifestation of the subject's unwillingness to accept unequal distributions and thereby a signal for the other members of a society

---

[5] Benign and invidious envy are standard distinctions in literature (though, as it will turn our below, I slightly modify the standard definition of invidious envy); the distinction of temperate envy is my own proposal. For a more detailed analysis of envy see, e.g., Schoeck (1969).

[6] The part in italics does not appear in the definitions of invidious envy. However, I think that it is necessary to introduce it in order to make clear the difference between invidious envy and what I call "temperate envy," which, in my view, are two distinct forms of envy.

that the agent will not accept distributions that fail to award her a reasonable part of a good being divided. Thus, temperate envy strengthens the subject's motivation to pursue the good she desires and constitutes a protection against the others' attempts to take advantage of her. One should examine a certain objection which can be raised against the above account of temperate envy. The objection says that this account implies that temperate envy presupposes some intuitions of justice, and that thereby it is a manifestation of some previously existing sense of justice, and not something that constitutes its rudimentary form. This objection can be refuted in the following way. Temperate envy is egocentric—aimed to protect the self-interest of the envious person. The sensitivity to unequal distributions built into temperate envy is not the sensitivity to unequal distributions *as such* but *as doing harm* to the envious person—it is therefore a manifestation of her concern with her own self-interest. In this sense temperate envy can be subsumed under the already mentioned concept of "disadvantageous inequality aversion."

The thesis that envy underlies our sense of justice is by no means novel—it has been advocated by many thinkers. For instance, Sigmund Freud held that our concern for equal treatment is the product of our childhood's envy; also Friedrich Nietzsche claimed that our moral intuitions arise from an emotion similar to envy, which he called "*ressentiment.*" From among contemporary thinkers who hold this view let me mention the renowned primatologist Frans B. M. de Waal. De Waal states that our noble principles have less noble origins and develops this view asserting that the sense of justice arises in the agent from her envy occasioned by receiving less than others and her anticipation of others' envy when she receives more than the others, and thereby from her willingness to avoid conflicts with them. De Waal's argument is, therefore, that our sense of justice arises from our envy and from our ability to predict others' envy (see de Wall 2006, 271-273). The claim that envy is an evolutionary adaptation is by no means novel. The only novel element in my account of the role of envy in the origins of our sense of justice seems to be the claim that envy is part of our rudimentary sense of justice (and not something out of which our rudimentary sense of justice arises), that this rudimentary sense of justice embraces other components, and that envy in question is temperate envy (and not, as most scholars seem to assume, invidious envy).

The instinct for retaliation

The rudimentary sense of justice not only determines our expectations about the share of a good we should receive but also propels us to punish those who violate those expectations, i.e., functions not only at the level of material justice, but also at the level of retributive justice. It is precisely the instinct for retaliation—a propensity to experience intensified anger issuing in vengeful or retaliatory actions—which underlies the rudimentary sense of retributive justice, i.e., motivates an agent who was treated in a way violating moral norms inflict the punishment on the perpetrators of this treatment.[7] It seems that the instinct for retaliation is a product of natural selection: generally speaking, it is evolutionary advantageous for a victim of immoral treatment to punish a person who violated moral norms with regard to her, as it shows the wrongdoer and other potential violators of these norms that the person who displays this emotion cannot be easily exploited. Accordingly, a person who is not psychologically disposed to punish those who cheated on her or otherwise violated moral norms with regard to her cannot be successful in cooperative interactions.

Gratitude

Evolutionary theory suggests that the emotion of gratitude arose in the context of the systems of reciprocity in order to support them (see Trivers 1983). It is therefore one of the motivational mechanisms upholding reciprocal altruism: an agent disposed to feel gratitude towards those who did her a favour is likely to reap higher gains from social exchanges than an agent who is not endowed with this disposition. This is so for three main reasons. First, gratitude motivates the agent to reciprocate, i.e., to refrain from defecting, and thereby serves upholding reciprocal exchanges.[8] One of the ways gratitude strengthens our motivation to reciprocate is by making us less sharp in discerning potential egoistic motives standing behind other people's kind actions towards us. It is clear that if we were aware of these motives, then our motivation to reciprocate would assuredly be weaker. Second, the reputation of a person disposed to experience this emotion makes her a desirable partner of reciprocal exchanges, and consequently is likely to widen the scope of reciprocal exchanges in which the person is involved. Third, gratitude, so to say, has a radiating character, i.e.,

---

[7] The instinct for retaliation is difficult to characterize in a precise way. Apart from the above characterization of this instinct—as a propensity to experience intensified anger issuing in vengeful or retaliatory actions, one may propose also a slightly different one (though not inconsistent)—as retaliatory spitefulness. The instinct of retaliation resembles pure spitefulness in that it moves an agent to sustain costs in order to make someone sustain even higher costs. However, retaliatory spitefulness essentially differs from pure spitefulness because while the latter is directed against the agents who did us no harm, the former is directed against those who did us harm. For an extensive treatment of the claim that our sense of justice is based on the transformation of the instinct for retaliation see Jacoby (1983).

[8] Michael E. McCullough and his colleagues label this function of gratitude "a moral motive function." See McCullough *et al.* (2001).

it moves an agent not only to reciprocate to those who did her a favour but also to initiate cooperative relationships—by doing favours—with agents with whom she has not so far kept such relationships. This "radiating" character of gratitude seems also favourable from the evolutionary perspective, as it engages us in new reciprocal exchanges and thereby opens to us the prospects of additional benefits. The capacity to feel the emotion of gratitude therefore serves in the long run our own interests. McCullough, Kilpatrick, Emmons, and Larson point also to another function of gratitude which they call "a moral reinforcer function:" gratitude of beneficiaries encourages benefactors to engage in further reciprocal exchanges in the future. Of course, this function also supports the system of reciprocity. One can briefly summarize the above remarks in game-theoretic parlance by saying that gratitude brings evolutionary advantages to those capable of feeling it, as it (like, in fact, all the other emotions discussed in this section) supports the tit-for-tat strategy, which proves to be especially efficient in promoting cooperation, and thereby supports the systems of reciprocity.

The analysis of gratitude would be incomplete without making some additional remarks. First, actions motivated by gratitude may be sometimes irrational (the beneficiary of a favour may be moved by gratitude to reciprocate even though the probability of further interactions with a partner who did her a favour is small as well as the probability of detecting her defection by the other members of a group). However, as mentioned, overall, gratitude is a fitness-maximizing emotion. Second, the intensity of gratitude is likely to depend on the value of received benefits (the higher the benefits are, the more intense gratitude is likely to be), on what intentions the beneficiary ascribes to the benefactor (if the benefactor is perceived as acting on egoistic motives or unintentionally, gratitude—if at all present—will be less intense), and on whether these benefits were perceived as due or not (in the former case gratitude will be less intense). Third, gratitude should be distinguished from another affective reaction to receiving benefits—a feeling of indebtedness. While gratitude is a pleasant emotion, which motivates people to reciprocate as a means of expressing one's good feelings toward the benefactor, a feeling of indebtedness is an unpleasant emotion, which motivates people to reciprocate as a means of reducing aversive arousal (see Greenberg 1980). Fourth, I would like to stress that the above account of gratitude refers only to its simplest form. Undoubtedly there are more complex forms of gratitude, e.g., the form of gratitude we feel toward those who exhibited to us a truly disinterested kindness, benevolence, good-heartedness, or who made us gifts (e.g. the gift of life) which cannot be returned. These more complex forms of gratitude are not the elements of the rudimentary sense of justice but very sophisticated ethical emotions.

The sense of guilt

As Robert Trivers wrote:

> It seems plausible [...] that the emotion of guilt has been selected for in humans partly in order to motivate the cheater to compensate his misdeed and to behave reciprocally in the future, and thus to prevent the rupture of reciprocal relationships. (Trivers 1971, 50)

It seems therefore that sense of guilt has arisen in the context of reciprocal altruism as an agent's subjective response to her failure to act cooperatively. Accordingly, it seems that the original context in which we feel a sense of guilt is when we fail reciprocate to those who have acted cooperatively towards us. The sense of guilt felt *ex post* motivates an agent to undertake reparative actions and thereby enables her to be re-involved in reciprocal relationships from which she would otherwise be excluded. The anticipated sense of guilt may, in turn, prevent an agent from defecting in the first place. Furthermore, if an agent is known to possess the sense of guilt, she is likely to be chosen for reciprocal exchanges. Thus, the sense of guilt seems to be a fitness-enhancing emotion.[9] Clearly, the above account of the sense of guilt does not exhaust this phenomenon—it refers only to its simplest form.

Digression

An additional argument for the claim that the rudimentary sense of justice is a biological adaptation comes from the animal studies, especially the studies of nonhuman primates. The studies show, among other things, that some primates reject unequal distributions of goods and punish those individuals who cheat in reciprocal exchanges.[10] It seems that these studies can be interpreted as attesting the fact that nonhuman primates display some primitive form of the sense of justice whose core is some kind of aversion to inequality. I call this form of the sense of justice "primitive" for three reasons. First, it is egocentric, not impartial: an ape manifests it only (or, at any rate, mainly—the data are not unambiguous here) when *she* is badly treated (i.e., receives less than other apes), not when *other apes* are badly treated (i.e., receive less than this monkey). An ape's sense of justice is therefore in the first place her expectation of

---

[9] For a further analysis of the sense of guilt as a fitness-enhancing emotion see also, e.g., Trivers (1985) and Ketelaar (2004).

[10] Experiments show, e.g., that a capuchin rejects a cucumber as a reward when she sees that another capuchin is offered a grape (the good that is more valued by capuchins than cucumbers); see, e.g., Brosnan (2006).

how *she* should be treated, and only marginally (if at all) her expectation of how *also others* should be treated. Technically speaking, apes manifest "disadvantageous inequity aversion" (they react when another individual receives a superior reward), not "advantageous inequity aversion" (they do not react when they receive a superior reward) (see Brosnan 2006). Second, as it seems, apes do not have a clear awareness of the content of the norms that they implicitly assume, i.e., which are encoded in their expectations–their sense of justice lacks therefore the cognitive element (perhaps they do not possess any awareness of the content of the norms). Third, the expectations of apes are not nuanced–they are a far cry from sophisticated norms of justice implied by the genuine sense of justice. It should be added, though, that this primitive sense of justice displayed by primates is in many respects similar to the rudimentary sense of justice of human beings, which, as was mentioned, is not fully impartial, does not necessarily presuppose the clear awareness of the content of the norms of justice (though it always presupposes *some* awareness of this content), and the norms of justice it implies may not be very sophisticated. The studies of primates show also that they exhibit emotions similar to those which are considered as an element of the rudimentary sense of justice: envy, the instinct for retaliation, gratitude and the sense of guilt.[11]

In sum, the above considerations lead to the following theses about the rudimentary sense of justice:

(1) The rudimentary sense of justice is constrained greed coupled with a bundle of emotions (envy, the instinct for retaliation, gratitude, the sense of guilt). The claim that the rudimentary sense of justice is constrained greed coupled with a bundle of emotions implies that the rudimentary sense of justice is not a phenomenon *sui generis*, as it can be decomposed into its simpler, constituent elements.

(2) The rudimentary sense of justice is a biological adaptation: those who were endowed with it had higher chances of survival and reproductive success than those who were not. It has evolved above all in the context of the relations of social exchange (reciprocal altruism); one of its basic functions seems to be a defence against exploitation in this context. The basic function of the rudimentary sense of justice, then, is to support the tit-for-tat strategy.

(3) All the differences between the above mentioned emotions notwithstanding, they fulfil two common general functions–they are motives to actions and signals to the others: they

---

[11] For more information on the primates' sense of justice see, e.g., De Waal (2006).

motivate those who are equipped with them to undertake actions that increase their chances of reaping high benefits from cooperative interactions, and they signal to others that those who are equipped with them are reliable and non-exploitable participants of cooperative interactions.

(4) As it seems, the rudimentary sense of justice implies or generates the following simple principles of justice: the egalitarian and perhaps also simple meritorious principles in the context of distributive justice and exchange-regulating justice (the latter context can also be called "the context of reciprocal altruism"), the principles "to each such a punishment which corresponds to the level of her guilt and the level of harm caused by her action" or "establish the just state of affairs" in the context of retributive justice. Thus, since the rudimentary sense of justice implies or generates certain normative meanings of justice, it can be said that it is through this form of our sense of justice that humans gain first insights into the possible normative meanings of justice. It should be remembered, though, that, at this stage of the development of their sense of justice, agents accept these principles mainly for egoistic reasons and often only dimly realize their content. The moral limitation of the rudimentary sense of justice can be clearly seen when we ask if a person endowed with this form of the sense of justice will be inclined to reject a distribution of a good which awards her (without any good reasons) more of this good than it does other persons, and which exposes her to no long-run negative consequences. It is obvious that constrained greed and envy will not make her reject this distribution. Greed, i.e., willingness to get as much as possible for oneself, can be constrained by the capacity to anticipate reactions of other people, but in the analysed case reason does not recommend accepting any constraints, as no negative consequences for the subject are likely to appear as a result of her accepting the offer. As for envy, it cannot, by definition, be experienced in situations in which inequalities in the distribution of a good favour the subject. This limitation of the rudimentary sense of justice is mitigated to some extent by the fact that this sense of justice embraces not only negative emotions but also a noble one–gratitude.

5 From the Rudimentary to the Genuine Sense of Justice

An interesting question is how the transition between the rudimentary and the genuine form of our sense of justice can come about. I shall present two competing accounts of this transition. The first one assumes that there is continuity between both forms of the sense of justice, the second one assumes that there is discontinuity between them. However, before I turn to discussing those accounts, I would like to articulate, at the risk of repeating some points made earlier, the differences between the rudimentary and genuine sense of justice.

Now, the question is how the genuine sense of justice can arise. It is implausible to maintain that the genuine sense of justice is a biological adaptation. Evolution has provided us with a truncated form of the genuine sense of justice, i.e., with the rudimentary sense of justice. For the rudimentary sense of justice to develop into the genuine sense of justice social learning is necessary. It is not clear, though, whether it is also sufficient. One may argue that what is also needed for this transition between the rudimentary and the genuine sense of justice to take place, is what may be called "a radical transformation of heart," or, to put it less poetically, a radical transformation of one's motivational structure which consist in decided and durable rising above one's natural self-absorption. The results of such a transformation is that an agent not only starts seeing clearly the nuanced principles of justice but also becomes capable of being motivated by those principles alone and thereby capable of overcoming her greed, the instinct for retaliation, and envy. We have therefore two accounts of the transition between the rudimentary and genuine sense of justice. The first one assumes that there is continuity in this transition. It can be presented as follows:

the rudimentary sense of justice + social learning → the genuine sense of justice

The second one assumes that this transition involves a moment of discontinuity (the moment is the radical transformation of heart). It can be presented as follows:

the rudimentary sense of justice + social learning + *the radical transformation of heart* → the genuine sense of justice

It is difficult to say which of these accounts is correct. My conjecture is that social learning can transform the rudimentary form of the sense of justice only up to a point in which an agent starts to see clearly nuanced principles of justice but it cannot radically change her motivational structure, i.e., it cannot engender in her the capacity to be motivated by the principles of justice alone. Accordingly, it may give rise to a sense of justice that can be situated between its rudimentary and genuine form: this intermediate form of the sense of justice would have, on the one hand, a distinctly developed cognitive element (like in the genuine form) but, on the other hand, its motivational element would be still based on constrained greed and the evolved emotions (like in the rudimentary form). The step necessary to attain the level of the genuine sense of justice would be the above mentioned radical transformation of heart. One may hold, of course, the view that people never reach the level of the genuine sense of justice, that the radical transformation of heart is always nothing more than a sham and fiction. This pessimistic—but not very plausible—view of our moral capacities implies that constrained greed and the evolved emotions not only constitute the rudimentary form of our sense of justice but also always constitute true motives of our just actions.

References

D'Arms, J. (2008). Envy, *The Stanford Encyclopedia of Philosophy*, http://plato.stanford.edu.
Brosnan, S. F. (2006). Nonhuman Species' Reactions to Iniquity and their Implications for Fairness. *Social Justice Research* 19 (2): 153-185.
De Waal, F. B. M. (2006). *La scimmia che siamo* (*Our Inner Ape*), transl. F. Conte. Milano: Garzanti.
Greenberg, J. (1980). A Theory of Indebtedness. In K. Gergen, M. S. Greenberg and R. H. Willis (eds.), *Social Exchange: Advances in Theory and Research*. New York: Plenum, 3-26
Jacoby, S. (1983). *Wild Justice: The Evolution of Revenge*. New York: Harper & Row.
Ketelaar, T. (2004). Ancestral Emotions, Current Decisions: Using Evolutionary Game Theory to Explore the role of Emotions in Decision Making. In *Evolutionary Psychology, Public Policy and Personal Decisions*. C. Crawford and C. Salmon (eds.). Mahwah: Lawrence Erlbaum Associates, 145-168.
McCullough, M. E. *et al.* (2001), Is Gratitude a Moral Affect? *Psychological Bulletin* 127 (2): 249-266.
Schoeck, H. (1969). *Envy. A Theory of Social Behaviour*. Indianapolis, Ind.: Liberty Press.
Trivers, R. L. (1971). The Evolution of Reciprocal Altruism. *Quarterly Review of Biology* 46.
_____ (1983). The Evolution of a Sense of Fairness. In *Absolute Values and the Creation of the New World: Proceedings of the Eleventh International Conference on the Unity of Sciences*. New York: International Cultural Foundation Press.
_____ (1985). *Social Evolution*. Menlo Park, Cal.: Benjamin/Cummings.
Załuski, W. (2009). *Evolutionary Theory and Legal Philosophy*. Cheltenham: Edward Elgar 2009.

19

Evolution, Religion, and the Human Mind

Slawomir Sztajer

In this chapter I would like to focus on a new cognitive science of religion, with a special consideration of the evolutionary perspective on religion. First, I present a general outline of the cognitive science of religion and its relation to the former approaches in the study of religion. Then, I consider some evolutionary scenarios concerning religion. Finally, I will try to list several philosophical consequences of this approach. In the whole chapter I am trying to point out that the study of religion is currently being revolutionised by the cognitive and evolutionary accounts.

Cognitive science is an interdisciplinary approach studying the cognitive system. It focuses on cognitive processes, the mind and intelligence, and includes such disciplines as psychology (cognitive and evolutionary), philosophy, neuroscience, artificial intelligence and cognitive anthropology. The fundamental thesis of cognitive science states that thinking can be understood in terms of representational structures in the human mind and the computational processes that operate upon these structures. Such a computational-representational conception of the mind turned out to be quite inspiring both theoretically or experimentally. It resulted in the construction and testing of various models of the mind. The results of this research have recently been applied to explanations of social and cultural phenomena.

Cognitive science is relatively new and dates back to the 1950s. The institutionalisation of cognitive science began in the 1970s, when the Cognitive Science Society was established and the first issue of the *Cognitive Science* journal was published. Today, cognitive science is being developed at many universities and research centres, mainly in Europe and North America. This development involves not only the creation of extended research programmes, but also the establishment of cognitive science study programmes.

In the 1990s the achievements of cognitive science were applied to the study of religion. Amongst the pioneers of this new paradigm were such scientists as Pascal Boyer, Stewart Guthrie, Thomas Lawson, Robert McCauley and Harvey Whitehouse. Since that time many academic publications in the field of the cognitive science of religion have appeared. Today, apart from the aforementioned pioneers, there are many other scientists who play a crucial role within the field, including such researchers as Ilkka Pyysiäinen, Justin Barrett, Jesse Bering, and others. They are, first and foremost, psychologists and anthropologists. Cooperation in the field of the cognitive science of religion has resulted in an establishment of the International Association for the Cognitive Science of Religion, and in publishing a series of publications in the *Journal of Cognition and Culture*, and in many other academic journals.

There is no single cognitive theory of religion and no precisely defined field of research within the cognitive science of religion. There is, however, advanced cooperation and an exchange of ideas. There is also a set of key issues that has absorbed cognitive scientists. The most important issues are the following questions: "1) How do human minds *represent* religious ideas? 2) How do human minds *acquire* religious ideas? 3) What forms of *action* do such ideas precipitate?" (Lawson 2000, 344)

1 Criticism of the Traditional Study of Religion

Cognitive scientists of religion criticise the traditional study of religion by focusing on the following issues. Firstly, they defy the conception of religion as a *sui generis* phenomenon. Their position may be described as (ontological and epistemological) reductionism. However, there is a difference between traditional reductionist theories of religion and the theories presented within the framework of the cognitive science. Whereas the latter avoids the monocausal explanation, the former usually simplifies religion by reducing it to one of many factors. Secondly, they criticise the traditional study of religion for ignoring aspects common to religion and to other areas of human activity. They claim that ordinary cognitive processes, which religion shares with other cultural forms, are fundamental to religion. Moreover, these processes are natural because they are the result of human adaptation to the environment. Thirdly, cognitive scientists of religion reject both an exclusively hermeneutic and exclusively explanative approach to religious phenomena and call for a balance between these two approaches (Lawson and McCauley 2006; Pyysiäinen 2003). At the same time, they emphasise the fact that the former science of religion has focused more on interpretation than explanation. This attempt to overcome one-sidedness of the traditional science of religion seems to be an attempt to bridge the methodological gulf between the natural sciences and the humanities. Fourthly, one of the fundamental mistakes of the science of religion, especially the anthropology of religion, was to study religious ideas from an epistemic rather than cognitive point of view. Studying religious ideas from the epistemic point of view means that they are apprehended in their relation to the world and presented as elements of the knowledge of the world. This results in raising the question about the truthfulness of these ideas. In contrast, studying religious ideas from the cognitive point of view means that the students of religion ask why people have certain ideas and ignore the question whether these ideas represent something in the world or not (Boyer 1994, 50). The cognitive science of religion studies religious representations in terms of causes rather than reasons. Fifthly, the cognitive science of religion criticises the tendency to accentuate differentiation and incomparability of religious phenomena, which results in ignoring the universal dimension of religion. It does not limit itself to making the statement that there is a religious pluralism but aims to find the common elements present in all religions.

2 Selected Achievements of the Cognitive Science of Religion

Among the cognitive conceptions of religion, the following seem to be of fundamental importance: (1) The naturalness of religion thesis; (2) The principle of "minimal counter-intuitiveness" of religious ideas (Pascal Boyer); (3) The principle of "theological correctness" (Justin Barrett); (4) The conception of religion as a kind of anthropomorphism (Stewart Guthrie); and (5) The cognitive theory of ritual (E. Thomas Lawson, Robert M. McCauley). It must be emphasised, however, that the aforementioned conceptions are not shared by all representatives of the new discipline, and that they are not treated as indisputable truths. On the contrary, with the exception of the first one, all of these issues are widely discussed today within the disciplinary field.

3 The Naturalness of Religious Ideas

The naturalness of religious ideas can be understood within the framework of the cognitive science of religion in two interrelated ways. Firstly, it means that religious ideas are ubiquitous. What is to be explained is not the presence of religious ideas in a given society but rather the fact that there might not have been any such ideas in some societies. As Justin Barrett says, "Being an atheist is not easy. In many ways it just goes against the grain. As odd as it sounds, it is not natural to reject all supernatural agents" (Barrett 2004, 108). Secondly, the naturalness of religious ideas simply implies that they are not supernatural and, as such, they can be explained in a purely natural way. This means that religion is not an extraordinary kind of human activity (experience, cognition); on the contrary, it is based on common psychological mechanisms. A student of religion should not assume that religion constitutes an activity that is fundamentally different from everyday life and that it involves some special cognitive abilities, experiences, unusual types of emotional commitment or brain states.

He or she should rather focus more on the natural foundations of religion. In this second sense, described by Pascal Boyer, natural are those aspects of religious ideas which depend on noncultural constraints, such as the human genome, abilities of the human brain, etc. (Boyer 1994, 3).

The content and structure of religious ideas depend on noncultural universal cognitive constraints which have their source in the evolution of human beings, especially in the evolution of the mind-brain. Cognitive scientists of religion often refer to naturalistic conceptions of the transmission of mental representations created by evolutionary psychologists and some representatives of sociobiology, as well as to other theories, such as Richard Dawkins' memetics or Dan Sperber's epidemiology of representations.

Representatives of the cognitive approach to religion assume that there are cognitive structures under-determined by experience. These "cognitive structures" orient the subject's attention to certain aspects of the available stimuli and narrow down the range of possible generalisations. Some aspects of these constraints cannot be directly derived from the experienced stimuli; on the contrary, they are a necessary condition if the experience in question is to have any cognitive effects at all (Boyer 1994, 26). This does not mean that these structures are inborn; it only means that they constrain the process of acquiring religious ideas. Human minds create a considerable number of religious ideas and countless variants of particular ideas. Some of them are more easily transmitted and acquired by individuals than others. Only those ideas which fulfill the requirements determined by cognitive structures are successfully spread.

4 The Evolutionary Point of View: Religion as an Adaptation and as a By-product of Adaptation

It is typical for the cognitive science of religion to search for the basis of religion in the functioning of the human mind, which is a product of a long-lasting process of evolution. This raises a very important question: To what extent are religious beliefs and actions adaptive? Does having religious beliefs favour human adaptation to the environment in some way or other? This question is part of a complex problem concerning the biological (evolutionary) basis of cultural phenomena and is a subject of discussion among scholars.

The evolutionary basis of religion can be understood in two ways: in the strongest version religion is a biological adaptation, in the weakest one it is a by-product of psychological mechanisms that had or still have an adaptive character. The majority of representatives of the cognitive science of religion claim that religion is a by-product of the evolved cognitive mechanisms. It must be emphasised here that, as Ilkka Pyysiäinen suggests, religion cannot be explained with reference to mechanisms that have evolved especially for the use of religious thinking. On the contrary, religion is possible thanks to various mental mechanisms that evolved as a solution to our Pleistocene ancestors' adaptive problems (Pyysiäinen 2004, 340).

In recent years there have been two main approaches studying culture from the evolutionary point of view, namely sociobiology and evolutionary psychology. The cognitive science of religion has been under the influence of evolutionary psychology and, thanks to it, it has been able to avoid an oversimplified conception of the relations between human biology and culture. Evolutionary psychologists have been more careful in identifying the biological determinants of culture. As far as the study of cultural phenomena is considered, cognitive psychology differs from sociobiology in the following respect:

> Sociobiologists tend to assume that the behaviors of humans in cultural environments are adaptive. They seek therefore to demonstrate the adaptiveness of cultural patterns of behavior and see such demonstrations as explanations of these cultural patterns. Evolutionary psychologists, on the other hand, consider that evolved adaptations, though of course adaptive in the ancestral environment, in which they evolved, need not be equally adaptive in a later cultural environment. Slowly evolving adaptations may have neutral or even maladaptive behavioral effects in a rapidly changing cultural environment. (Sperber and Hirschfeld 1999, cxiv)

For evolutionary psychologists the mind is a mediating element between genes and behaviour. Since the cognitive science of religion has been under the influence of evolutionary psychology, its representatives claim that religion is not an adaptation but rather a by-product of certain psychological mechanisms. Although some of these mechanisms might have been or still are adaptive, this is not true in reference to religious ideas that exist thanks to evolved cognitive mechanisms. According to this theory, the ubiquity of some religious ideas, especially those which are counter-intuitive representations of supernatural beings, is a product of definite mental abilities. Religion would thus be a by-product of ordinary cognitive processes.

5 The First Scenario: Religion as an Adaptation

This scenario is accepted by the neurologists Erica Harris and Patrick McNamara. They assume the existence of specific religious neuroanatomy and neurophysiology, and claim that religiousness satisfies the minimal criteria to be a biocultural adaptation. Since religion is conditioned by both biological and cultural factors, it is reasonable to talk about "biocultural" instead of mere "biological" adaptation.

According to Harris and McNamara, in order to be considered a biocultural adaptation, any human trait must satisfy at least three criteria: "(1) universality across cultures; (2) relative effortless-ness of acquisition of the trait (the trait is not merely learnt); (3) and an associated 'biology,' which refers to a consistent set of physiologic systems that reliably support, mediate and produce the trait or behavior in question" (Harris and McNamara 2008, 79–80). By "biology" they mean "(1) a genetic component as evidenced by genes-behavior correlations and heritability studies; (2) a brain component as evidenced by classical neuropsychology and neuroimaging studies; and (3) a chemistry component as evidenced by pharmacologic studies" (ibid., 80). The first two criteria of being a biocultural adaptation, i.e. universality across cultures and relative effortlessness of acquisition, seem to be satisfied in the case of religiousness. When religiousness is defined as a human practice involving beliefs in supernatural beings and ritual performances related to those beliefs, it is found all over the world. As many anthropologists suggest, it is very difficult, if not impossible, to find a human society without religion. Religious ideas and behaviours are considered to be human universals. They are also easily acquired. They are not just imposed on children by their religious parents and teachers, but are rather spontaneously acquired, and even created, by children.

The last requirement that has to be met in order for religiousness to be considered a biocultural adaptation is that physiological systems must exist which produce and support religious representations and behaviours. It is highly controversial whether religiousness satisfies this last criterion. The problem is that not much research has been done so far in this area and that it is difficult to find satisfying evidence for the existence of biological foundations of religiousness. Harris and McNamara claim that religiousness is indeed supported by a dedicated biological system. Firstly, they point out that religiousness is heritable to a considerable degree, which means that it has some genetic component. By analysing twin studies the authors came to the conclusion that in the case of religiousness, the heritability coefficient is moderate or even high. This means that there is a correlation between genes and religiousness. Secondly, on the basis of neuroimaging studies they claim that one can observe high activation in the right prefrontal cortex during religious activity. Thirdly, the authors insist that there is a correlation between dopamine transmission in the brain and religiousness. Religiousness, then, depends on the level of dopaminergic activity.

The three criteria constitute the necessary conditions that must be fulfilled in order for religiousness to be an adaptation. However, it does not suffice it to say that there is a specific religious neuroanatomy and neurophysiology. The adaptationist view usually involves questions about in what sense religiousness is adaptive and which adaptation problems it solves. For many representatives of the evolutionary approach, religion solves problems related to social interactions. By using reverse engineering, adaptationists claim that religion solves adaptive problems such as getting along with others, and getting along with ourselves (Bulbulia 2005, 81). Religion enhances cooperation between individuals, stabilises social exchange and limits rational motives for cheating. It renders social exchange more predictable and, in this way, maintains the existence of society. The above claim is connected with the assumption that biological success is better when individuals form a coalition with other individuals and face problems together. Religion is adaptive because it enhances the chances of survival by maintaining cooperation between unrelated individuals. The adaptive character of religion may also be connected with the role which religion plays in the inner life of individuals, i.e. of giving them meaningful, integrated and optimistic outlooks on life (Bulbulia 2005, 89).

Religiousness can be adaptive either on a group level or the individual level. In the first case it enhances the fitness of groups compared to other groups (between-group selection); in the second case, religion enhances the fitness of individuals compared to other individuals in the same group (within-group selection) (Wilson 2008, 24). According to David Sloan Wilson, there is also one other possibility: as far as cultural evolution is considered, it is possible to think of religion as a cultural parasite, i.e., a trait which evolved "to increase its own transmission, like a disease organism, without benefiting human individuals or groups" (Wilson 2008, 24).

6 The Second Scenario: Religion as a By-product

The second scenario of the evolutionary origin of religious beliefs and behaviours is based on the idea that religiousness is not directly adaptive, but is a by-product of traits, especially cognitive mechanisms which were either adaptive in the past or are still adaptive today in non-religious contexts. For example, religiousness could have been advantageous in the human ancestral environment, when people lived in small groups and faced many natural threats (e.g. the possibility of being attacked by various predators). Although today the ancestral conditions do not exist, the mental mechanisms that evolved to solve various adaptive problems still do and are responsible for generating religious representations and behaviours. By using the notion introduced by Stephen Jay Gould, one could say that religion is a spandrel, i.e. a secondary consequence of the evolution of some adaptive trait. Spandrels "arise nonadaptively as secondary consequences, but then become available for later cooptation to useful function in the subsequent history of an evolutionary lineage" (Gould 1997, 10750). This conception of religion is defended by such scholars as Steven Pinker, Pascal Boyer, Scott Atran, Lee A. Kirkpatrick, and many others.

Cognitive mechanisms were not designed to generate religious representations. They were rather co-opted for religious purposes. Among them are such mechanisms or tools as the Agency Detection Device and the Theory of Mind Mechanism. The cognitive mechanism described as the Agency Detection Device is a mental module which humans share with animals. ADD evolved because the detection of other organisms in the environment was essential for survival. In humans the device is hypersensitive: it is readily activated under the influence of a wide range of stimuli. Although from the point of view of survival and reproduction it pays to be hypersensitive to detecting agency, this strategy is costly because it often produces detection errors. The Agency Detection Device is responsible for generating the concepts of beings that are not physically present. It plays a crucial role in the forming of representations of a supernatural being. The other important mechanism is the Theory of Mind Mechanism, which is the human ability to attribute various mental states to real or supposed agents. On the one hand, the ability of mind-reading may be important for human survival and reproduction; on the other hand, it is responsible for attributing intentions to supposed agents. It seems that both of the cognitive mechanisms lie at the basis of the tendency to anthropomorphise the world. According to Steward Guthrie, religion is a kind of anthropomorphism which occurs under the influence of uncertainty. As the author writes, "uncertain of what we face, we bet on the most important possibility because if we are wrong we lose little and if we are right we gain much. Religion, asserting that the world is significantly humanlike, brings this strategy to its highest pitch" (Guthrie 1993, 38).

7 Philosophical Implications

The evolutionary approach to religion has several philosophical implications.

Firstly, a contemporary tendency to study religion as a natural phenomenon includes not only explicitly stated demands for more thorough concentration on the bio-psychological dimensions of this phenomenon, but also for tacit assumptions and ideological consequences which are manifestations of the philosophical interpretation of empirical research and theoretical speculations. In this context religion is not seen as a human dialogue with a transcendent reality, but rather as a purely human endeavour.

The evolutionary approach to religion is reductionist, which means that it aims at reducing religious phenomena to non-religious ones. From this point of view, religion seems to be a by-product of natural cognitive mechanisms. Each time this view of religion is confronted with the religious stance it challenges the religious worldview. It would be very difficult for a religious person to accept all of the consequences of this scientific approach. Cognitive and evolutionary approaches to religion differ from many traditional theoretical and methodological orientations developed in the study of religion in that they treat it as a purely natural phenomenon. Many traditional approaches have studied religion as a socio-cultural (Clifford Geertz) or transcendental phenomenon (e.g. some representatives of the phenomenology of religion, many philosophers of religion). But since there is nothing special in religious cognition, it can be explained in terms of natural phenomena.

The cognitive and evolutionary approaches marginalize the role of the religious consciousness in religion. It ceases to be a starting point in the study of religious phenomena. As a result, the usefulness of phenomenological and hermeneutical studies is being questioned. It is pointed out that the hermeneutical and phenomenological study of religious symbolism is nothing more but an interpretation extending, not explaining, of religious symbolism. Such an interpretation does not explain anything and sometimes even hampers the scientific study of religion. The possibility of applying natural science achievements to religions opens up a new perspective for the study of religion, thus establishing a balance between explanatory and interpretive approaches.

From the point of view of philosophy of religion, the cognitive science of religion has redefined the debate over the relation between religion and science. With the emergence of the naturalistic study of religion, the debate has lost its vitality. Religion is constantly seen less as a rival to science—it has become an object of scientific scrutiny. This does not mean that this long-lasting debate is now over. However, from the naturalistic perspective, the dialogue between religion and science would be equally as fruitless as the debate between science and art, or between science and common sense. This shift of emphasis places religion on the margin of contemporary culture, since it is no longer a partner in the debate, but rather a phenomenon that is to be reduced in explanation to a set of nonreligious factors. The scientific explanation of religion leads to the conclusion that religion presents itself as something other than it really is. The naturalistic explanation of religion is not the same as the explanation of any other cultural form. Since every religion postulates the existence of supernatural beings it is particularly endangered by scientific naturalisation. The supernatural claims of religion do not hold up to criticism.

It seems that the evolutionary perspective on religion marginalizes the question about the rationality of religion, especially the rationality of religious beliefs. In traditional philosophical debates, the rationality of religious beliefs has been connected to the issue concerning the truthfulness of these beliefs as well as the possibility of their justification. The naturalistic approach is not interested in reasons for having religious beliefs, but it does try to describe the origins and functioning of these beliefs—it is not important whether they are true or false.

References

Barrett, J. (2004). *Why Would Anyone Believe in God?* Walnut Creek, Cal.: AltaMira Press.

Boyer, P. (2001). *Religion Explained: The Human Instincts that Fashion Gods, Spirits and Ancestors.* London: Heinnemann.

Boyer, P. (1994). *The Naturalness of Religious Ideas: A Cognitive Theory of Religion.* Berkeley: University of California Press.

Bulbulia, J. (2005). Are There Any Religions? An Evolutionary Explanation. *Method and Theory in the Study of Religion* 17: 71-100.

Bulbulia, J. et al. (eds.) (2008). *The Evolution of Religion: Studies, Theories, and Critiques.* Santa Margarita, Cal.: Collins Foundation Press

Gould, S. J. (1997). The Exaptive Excellence of Spandrels as a Term and Prototype. *Proceedings of the National Academy of Sciences* 94:10750-10755.

Guthrie, S. (1993). *Faces in the Clouds: A New Theory of Religion.* New York: Oxford University Press.

Harris, E., and McNamara, P. (2008). Is Religiousness a Biocultural Adaptation? In Bulbulia (2008), 79-86.

Lawson, E. T. (2000). Towards a Cognitive Science of Religion. *Numen* 47: 338-349.

Lawson, E. T., and R. McCauley. (2006). Interpretation and Explanation: Problems and Promise in the Study of Religion. In *Religion and Cognition: a Reader.* D. J. Stone (ed.). London: Equinox.

Pyysiäinen, I. (2003). Gods, Genes, and Passion. *Journal of Cognition and Culture* 3: 175-185.

Pyysiäinen, I. (2004). Bridge over Troubled Water: Crossing Disciplinary Boundaries. A Review Essay. *Method and Theory in the Study of Religion* 16: 336-347.

Sperber, D. and Hirschfeld, L. (1999). Culture, Cognition, and Evolution. In *The MIT Encyclopedia of the Cognitive Sciences.* R. Wilson and F. Keil (eds.). Cambridge, Mass.: The MIT Press.

Sosis, R. (2009). The Adaptationist—By-product Debate on the Evolution of Religion: Five Misunderstandings of the Adaptationist Program. *Journal of Cognition and Culture* 9: 339-356.

Wilson, D. S. (2008). Evolution and Religion: The transformation of the Obvious. In Bulbulia (2008), 23-30.